Search for the Higgs Boson in the Vector Boson Fusion Channel at the ATLAS Detector

Eric Ouellette

Search for the Higgs Boson in the Vector Boson Fusion Channel at the ATLAS Detector

Eric Ouellette
Department of Physics
University of New Brunswick
Fredericton
New Brunswick
Canada

ISBN 978-3-319-36625-8 ISBN 978-3-319-13599-1 (eBook)
DOI 10.1007/978-3-319-13599-1

Springer Cham Heidelberg New York Dordrecht London

Softcover reprint of the hardcover 1st edition 2015

Printed on acid-free paper

Springer is part of Springer Science+Business Media (www.springer.com)

For my family.

Supervisory Committee

Dr. J. Albert, Supervisor
(Department of Physics & Astronomy)

Dr. R. Sobie, Departmental Member
(Department of Physics & Astronomy)

Dr. R. Keeler, Departmental Member
(Department of Physics & Astronomy)

Dr. A. Briggs, Outside Member
(Department of Chemistry)

Abstract

The search for the Higgs boson has been a cornerstone of the physics program at the Large Hadron Collider in Geneva Switzerland. The ATLAS experiment successfully discovered the Higgs using the so-called 'Golden Channels' of $H^0 \rightarrow \gamma$ and $H^0 \rightarrow ZZ^{(*)}$ using data samples collected during the 2011 and 2012 run periods. In order to check if the discovered Higgs is consistent with purely Standard Model behaviour, it is necessary to further confirm the existence of the Higgs in each production mode and decay channel predicted by the Standard Model.

For this dissertation, a search for the Higgs was conducted using the $H^0 \rightarrow b\overline{b}$ decay channel, where the Higgs is produced by the inverse pair decay of two weak bosons exchanged by a scattered quark pair, also known as Vector Boson Fusion (VBF). This analysis uses data samples collected during the 2011 run period by the ATLAS detector totalling 4.2 fb^{-1} of proton-proton collisions at $\sqrt{s} = 7$ TeV. No excess of events above background expectation is observed and 95 % confidence level upper limits on the Standard Model Higgs cross section times branching ratio, $\sigma(VBF) \times \mathrm{BR}\left(H^0 \rightarrow b\overline{b}\right)$, are derived for Higgs masses in the range $115 < m_H < 130$ GeV. An observed 95 % confidence level upper limit of 18.7 times the Standard Model cross section is obtained for a Higgs boson mass of 125 GeV.

Acknowledgements

Many people are to thank for all their support and guidance over the last few years. I'd like to thank to my family, for their endless support and encouragement, and always being a phone call away, despite being on the other side of the continent (or ocean). Thank you to the people I worked with in Geneva (Michele, Ricardo, Mathieu), for guiding me through my months at CERN, and always being willing to answer my questions. Thank you to everyone in the Department of Physics & Astronomy at UVic: the professors, the post-docs, the graduate students (past and present), and the support staff, for always being ready to lend a hand, in whatever problem I might have. A big thanks to all my friends, both new and old, past and present, here and abroad, who have always encouraged me to work hard, yet reminded me to enjoy life.

I would like to acknowledge the funding provided by NSERC, the University of Victoria, and the Department of Physics & Astronomy. A graduate student's ability to concentrate on their studies, and not their finances, is greatly underestimated, and the funding you provided certainly helped with my piece of mind while living in Victoria and (the ever expensive) Geneva.

Finally, I would like to thank my supervisor, Dr. Justin Albert, for all his help, his unending support, and his enthusiasm throughout my PhD. Your optimism and encouragement always made me believe that I could do it.

Contents

List of Figures

List of Tables

Acronyms

ATLAS	A Toroidal LHC ApparatuS
CMS	Compact Muon Solenoid
SM	Standard Model
EM	Electromagnetic
CERN	European Organization for Nuclear Research
LAr	Liquid Argon
LHC	Large Hadron Collider
QCD	Quantum Chromodynamics
QED	Quantum Electrodynamics
FCal	Forward Calorimeter
HLT	High Level Trigger
L1	Level 1
L2	Level 2
EF	Event Filter
RoI	Region of Interest
LB	Luminosity Block
GRL	Good Run List
LEP	Large Electron-Positron
VBF	Vector Boson Fusion
MC	Monte Carlo
PDF	Parton Distribution Function
TRF	Tagging Rate Function
NLO	Next-to-Leading Order
JES	Jet Energy Scale
CL	Confidence Level
CJV	Central Jet Veto

Chapter 1
Introduction

The elements that compose everything around us have been a constant source of curiosity for millenia, with our understanding of them continuously improving. We have come a long way since our classical belief in the existence of only four elements: earth, wind, water, and fire. Major progress was made in the nineteenth century, when Dimitri Mendeleev developed the Periodic Table of the Elements, a table that grouped similar elements together based on their chemical properties, and that was able to predict the existence of yet undiscovered elements.

Particle physics was arguably born from this Periodic Table, as scientists tried to delve deeper, and understand the differences between these elements. At the time, the atom was believed to be the smallest unit of matter, but this was proven false when, in 1897, J.J. Thomson discovered the electron by deflecting "cathode rays" using a magnet, and surmising that the ray was actually made up of particles of very small mass and of negative charge. Their existence implied that the atom was not fundamental, but rather made up of electrons floating in, what was believed at the time, a "soup" of positive charge.

Ernest Rutherford demonstrated that most of an atom's mass was concentrated in a tiny, positively charged core at the centre of the atom. This tiny core was called a "proton" for the hydrogen atom, believed to be the building block for all other elements. There was a discrepancy, however, when comparing the atomic mass and the number of protons per element. This was resolved in 1932 when James Chadwick discovered the "neutron", an electrically neutral twin to the proton. He determined that each atom's nucleus contains roughly the same number of neutrons as protons. With this, the building blocks of the atom were set: the proton, the neutron and the electron.

But, new particles continued to be discovered in the years following the discovery of the neutron: positrons (the antiparticle of the electron), muons, pions, kaons, and neutrinos. It was later discovered that protons and neutrons were actually not fundamental but rather made up of smaller constituents, called quarks and gluons. Each newly discovered particle appeared to have their own unique properties; however, some shared similarities to other previously discovered particles, which hinted at an underlying theory connecting them all.

E. Ouellette, *Search for the Higgs Boson in the Vector Boson Fusion Channel at the ATLAS Detector*, DOI 10.1007/978-3-319-13599-1_1

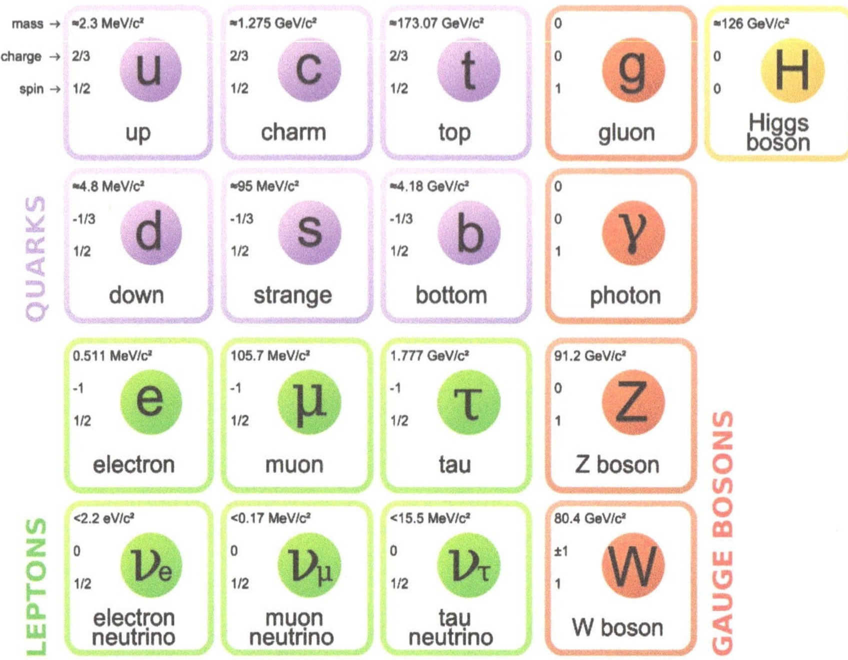

Fig. 1.1 The particles of the Standard Model of particle physics. Fermions are divided into two types: leptons and quarks. These are each organized into three pairs (or generations) of increasing mass. Four bosons are responsible for mediating interactions between the fermions via the three main forces of the SM: electromagnetic, strong and weak forces. The Higgs boson is responsible for giving mass to all other particles [4]

In the 1960's, theorists, inspired by the results from particle physics experiments, tried to explain the nature of these particles, and derive a model that fully described the interactions between them. The result is the Standard Model (SM) [1, 2, 3] of particle physics. Developed over several years, the model uses quantum field theory, a theoretical framework that treats particles as excited states of an underlying field, to describe particle interactions. Within the SM, all particles are categorized into one of two types: fermions or bosons. Fermions are further subdivided into two categories: leptons and quarks, with each category containing six particles and six antiparticles. Fermions are usually associated with matter, as atoms are composed of them: the electron (a lepton) and two quark-types that make up protons and neutrons. Bosons are particles responsible for mediating the interactions that occur between fermions via one of three fundamental forces: the electromagnetic force (responsible for the Coulomb force), the strong force (responsible for the binding of protons and neutrons within the nucleus) and the weak force (responsible for radioactive decay). The bosons associated to these forces are the photon, gluon and weak bosons ($W^{\pm}$ and Z^0), respectively. One additional boson also exists within the SM, the Higgs, which is responsible for giving mass to all other particles. A table summarizing the particles of the SM is shown in Fig. 1.1.

The SM is a modern-day Periodic Table, since it was able to predict the existence of yet undiscovered particles. For example, the W and Z bosons (of the weak force) were only discovered in the 1980's, whereas the top quark (the heaviest of all known particles) was only discovered in 1995. Prior to 2012, the Higgs boson was the last undiscovered particle of the SM, with scientists searching for this particle since first being theorized over 40 years ago. The SM does not predict the mass of the Higgs, making its discovery more difficult.

There are indirect methods to determine the mass of the Higgs; however, its very weak dependence on other observables made it very difficult to measure experimentally. Thus, direct searches were conducted for the Higgs, first by the Large Electron-Positron (LEP) collider in the 1990's and then by the Tevatron collider in the 2000's. Detectors at both colliders were unable to find the Higgs, but they were able to constrain the mass region where the Higgs could exist.

The Large Hadron Collider (LHC) was built in the tunnel previously used by LEP at the European Organization for Nuclear Research (CERN) in Geneva Switzerland in the 2000's, with the hope to find this elusive particle. Since it began operations in 2008, the experiments of the LHC were able to confirm the discovery of the sought after Higgs boson using data from 2011 and 2012.

The Higgs boson can be produced at the LHC through one of several mechanisms; it can also decay to a number of final states. The analyses used for the discovery of the Higgs concentrated on a small subset of these search channels, the so-called 'Golden Channels', due to their more manageable backgrounds and the experiment's ability to accurately measure the particles in their final states. To fully validate the SM, we need to find the Higgs in each production mode and decay channel combination predicted by the SM, as deviations in any of these could suggest physics beyond the SM. Searching for the Higgs in each different channel can also further help determine its properties, by measuring its coupling to all particles and comparing it to SM predictions.

The research for this dissertation involves a search for the Higgs in a new channel at the ATLAS experiment at the LHC. The Vector Boson Fusion (VBF) production mode of the Higgs, which is defined as the generation of a Higgs by the inverse pair decay of two weak (or vector) bosons exchanged by a scattered quark pair, is used. The decay of the Higgs to two b-quarks (the second heaviest of the quarks) is used as the decay channel. A schematic view of the search channel is plotted in Fig. 1.2. The VBF production mode is the second most abundant mechanism to produce a Higgs at the LHC; the $H^0 \rightarrow b\bar{b}$ channel is the most abundant decay mode for the Higgs mass discovered by the LHC experiments. The VBF production mode is also sensitive to other properties of the Higgs boson, such as spin and parity. As no other groups at ATLAS had attempted a search using this channel, the goal of this dissertation was to develop the first selection algorithm of the Higgs in this particular production and decay mode.

Finding the Higgs in this channel is difficult due to the background from other events. As the schematic in Fig. 1.2 clearly shows, the final state of this search channel is four quarks, two of which are b's. Since the LHC collides protons together, and protons are composed of quarks, the vast majority of events at the LHC have

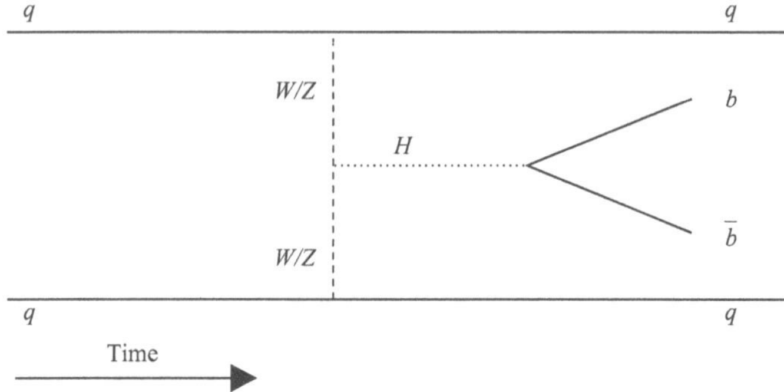

Fig. 1.2 Schematic diagram of the Higgs search channel used for this dissertation, VBF $H^0 \to b\bar{b}$. Note that in this schematic, time moves forward along the horizontal

similar final states. In order to develop an analysis strategy for this channel, one needs to effectively record relevant events during data taking and effectively reduce background while keeping as many signal events as possible.

In this thesis, the background theory needed to understand this topic (Chap. 2), and the status of the Higgs search and discovery (Chap. 3) are introduced. Then, a description of the LHC, ATLAS, and the facilities used for this experiment follows (Chap. 4). A thorough description of the analysis developed for this dissertation (Chaps. 5, 6 and 7) is next. Finally, results and a discussion (Chap. 8) conclude the dissertation.

Chapter 2
Theory

This chapter introduces the Standard Model (SM), and a brief overview of proton-proton collisions. The most common unit used throughout this dissertation is the GeV (the giga-electronvolt), which corresponds to the kinetic energy of an electron, accelerated through 10^9 V. Units of momentum and mass can be represented as GeV/c and GeV/c^2 respectively, with c the speed of light. To simplify notation, in this chapter and others, natural units are used, corresponding to $c = 1$. This allows energy, momentum and mass to all be measured in the same units, GeV[1], where 1 GeV approximates to the mass of a single proton or neutron (their masses are actually 0.938 and 0.940 GeV respectively).

2.1 Standard Model

The SM of particle physics is a theory that describes the currently known smallest sub-atomic particles and the interactions between them. In the SM, matter is made up of twelve half-integer spin particles called fermions, detailed in Table 2.1. Fundamental fermions can be classified into one of two categories: leptons or quarks. The fundamental fermions are further divided into three generations, with particle masses generally increasing from one generation to the next[2]. For each generation of lepton, there exists one particle with an electric charge of $-e$[3] and one neutral partner, called a neutrino. For each generation of quark, there exists one particle with an electric charge of $+2/3e$ and another with charge $-1/3e$. Each fermion has a partner with identical mass but inverse charge called an antiparticle. Leptons are

---

[1] The advantage of this can be seen when considering Einstein's famous equation $E = mc^2$. A particle of rest mass $m = 1$ GeV$/c^2$ has a rest energy of $E = 1$ GeV.

[2] With the exception of neutrinos, which are known to have very small yet unknown absolute masses.

[3] In particle physics, when dealing with charge, it is common to work in units of fundamental charge, e, which has as value 1.602×10^{-19} C.

E. Ouellette, *Search for the Higgs Boson in the Vector Boson Fusion Channel at the ATLAS Detector*, DOI 10.1007/978-3-319-13599-1_2

Table 2.1 Fermions of the SM, taken from the particle data group summary tables [5]

Generation	Leptons			Quarks		
		Charge [e]	Mass [GeV]		Charge [e]	Mass [GeV]
First	Electron, e	-1	5.11×10^{-4}	Up, u	$+2/3$	≈ 0.002
	e neutrino, ν_e	0	$< 2 \times 10^{-9}$	Down, d	$-1/3$	≈ 0.005
Second	Muon, μ	-1	0.1057	Charm, c	$+2/3$	1.3
	μ neutrino, ν_μ	0	$< 1.9 \times 10^{-4}$	Strange, s	$-1/3$	0.1
Third	Tau, τ	-1	1.777	Bottom, b	$+2/3$	4.2
	τ neutrino, ν_τ	0	$< 1.8 \times 10^{-2}$	Top, t	$-1/3$	173

Table 2.2 Bosons of the SM, taken from the particle data group summary tables [5]

	Mechanism	Boson	Charge [e]	Mass [GeV]
Spin-1	EM force	Photon, γ	0	0
	Strong force	Gluon, g	0	0
		$W^\pm$	± 1	80.4
	Weak force	Z^0	0	91.2
Spin-0	Higgs	Higgs, H^0	0	126

free to exist in nature by themselves, but quarks cannot. Instead they must combine together with other quarks to form hadrons. There exists two types of hadrons: baryons, comprised of three (anti-)quarks and having integer spin (i.e. qqq or $\bar{q}\bar{q}\bar{q}$), and mesons, comprised of a quark-antiquark pair and having half integer spin (i.e. $q\bar{q}$). The matter in our Universe is made up of first generation fermions, since atoms are made of electrons orbiting around a nucleus. Nuclei are composed of protons and neutrons, both of which are baryons, with the proton containing two u-quarks and one d-quark (uud) and the neutron containing two d-quarks and one u-quark (udd).

The SM includes five integer spin bosons, detailed in Table 2.2. Interactions between particles are mediated by four spin-1 bosons, so called "force-carrying" particles, called gauge bosons. These are: the photon (γ), the gluon (g) and the two weak bosons ($W^\pm$ and Z^0), which mediate the electromagnetic (EM), the strong and the weak force, respectively. Each force in turn has its own physical theory that describes how these bosons mediate the interactions between various particles. They are all built from a theoretical framework called quantum field theory, which treats particles as excited states of an underlying quantized field. One spin-0 boson is also included in the SM: the Higgs. It is responsible for giving mass to other particles via the Higgs mechanism. The underlying theories that describe the behaviour of these bosons are introduced briefly below [6–8].

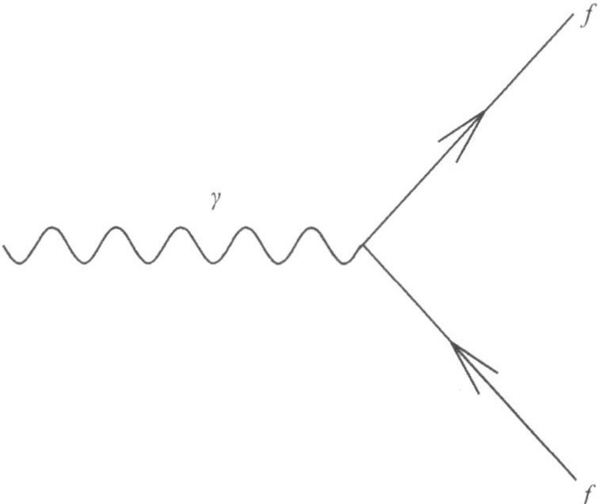

Fig. 2.1 Leading order QED vertex used in Feynman diagrams, where γ is a photon and f is any charged fermion

2.1.1 Quantum Electrodynamics

Quantum Electrodynamics (QED) is the relativistic quantum field theory that describes the interactions between particles that have electric charge. The force, which is mediated by the chargeless photon, has infinite range, but its strength is proportional to r^{-2}.

All interactions between particles are schematically described using Feynman diagrams (described briefly in Sect. 2.1.5) where one builds up particle interactions using the small set of allowable vertices. The leading order vertex for QED is shown in Fig. 2.1. Each vertex has a coupling constant that is proportional to the fine structure constant, $\alpha \approx 1/137$. The term constant is a bit of a misnomer, as it is slightly dependent on the momentum transfer of an interaction. QED is incredibly well understood, and has been extensively tested, with great success.

2.1.2 Quantum Chromodynamics

Quantum Chromodynamics (QCD) is the quantum field theory that describes the strong force, which are interactions involving partons (quarks and gluons). Similar to electric charge, quarks have a "colour charge", which can be one of three values (red, green or blue), and antiquarks have an anti-colour equivalent. Unlike the photon, which does not carry electric charge, the gluon has both a colour and an anti-colour. This implies two unique properties of the strong force: quark colours change when interacting with the gluon, and gluons can interact with itself. Leading order vertices of QCD are shown in Fig. 2.2.

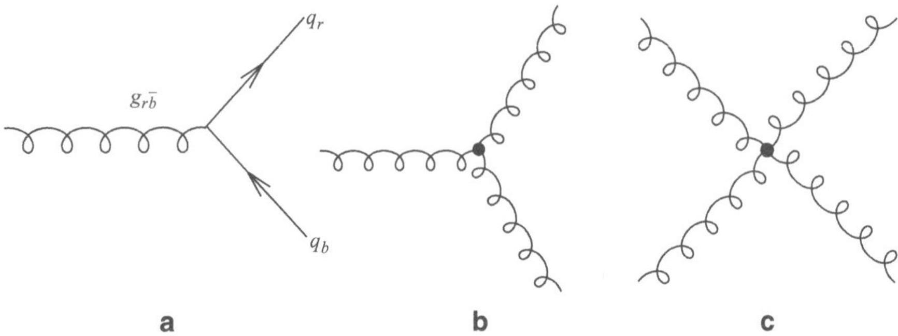

Fig. 2.2 Leading order QCD vertices used in Feynman diagrams: **a** the gluon-quark-quark interaction (note the colour of the gluon, g, is a simplified view of its actual colour), **b** the three-gluon self interaction, and **c** the four-gluon self interaction

Like the photon, gluons are massless, suggesting they can mediate a force of infinite range. However, colour charges are affected by a phenomenon known as confinement, which states that colour charged particles (quarks and gluons) cannot be isolated singularly, and must therefore always be bound in some way in a colourless configuration. These combinations were introduced above: mesons ($q\bar{q}$ where the colour of q is cancelled by the anti-colour of $\bar{q}$) and baryons (qqq where each q has a different colour, thus making it "white"). Due to confinement, when two bound quarks begin to separate, the strong interaction between them (mediated by the gluon) actually becomes stronger. The potential between the two quarks increases until there is sufficient energy to create a quark anti-quark pair. For this reason, the strong force is effectively a short range force.

Another particularity of QCD is that the force between quarks becomes weaker as the distance between them decreases or as the quark energies increase. This is caused by an effective "antiscreening" of colour charge as quark distances decrease. This phenomenon is called asymptotic freedom, and is essential in understanding proton-proton collisions (like those of the Large Hadron Collider (LHC). This also illustrates why protons and neutrons are able to be bound together in such a small space (the nucleus), but not repel each other despite having like electric charge.

As the name suggests, this force is stronger than the EM force, which is most evident when comparing the coupling constant for strong interactions, α_S, to that of EM interactions, α. However, the value of α_s is not constant and is dependent on the energy scale of the interaction, as evidenced by the plot in Fig. 2.3. The strong coupling constant α_s tends to large values at very low energy scales (or inversely, very large length scales), which demonstrates confinement and why quarks cannot exist singularly. Conversely, as energy scales increase (or length scales decrease), α_s decreases, which demonstrates asymptotic freedom and why quarks and gluons within a high energy proton can be treated as free particles.

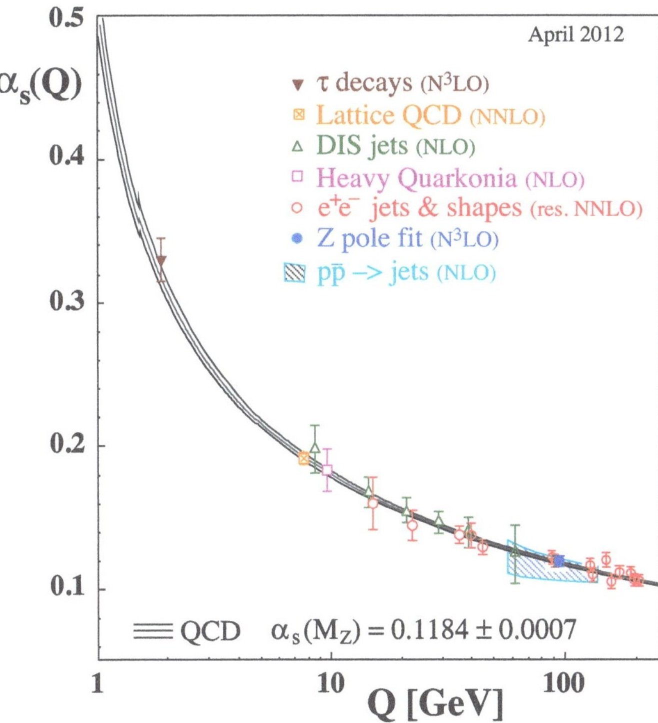

Fig. 2.3 Plot of the strong coupling constant, α_s, as a function of the energy scale, Q. The energy scale can also be considered the inverse length scale: as Q increases, the length scale decreases, and vice versa. Taken from the Particle Data Group reviews [5]

2.1.3 Weak Interaction

The weak interaction is different from both QED and QCD in that the mediating particles, the W and Z bosons, have mass, and the W has either positive or negative electric charge. The weak force is also unique in that quark flavour is not necessarily conserved in an interaction (ie. one quark can change from one flavour to another), and the parity and charge parity symmetries are violated.

At low energy scales, this force is much weaker than the EM and strong forces. The "weak coupling constant" can actually be regarded as greater than that of QED, with $\alpha_W \approx 1/30$; however, the strength of an interaction is suppressed by the massive mediating boson at low energies. As energy scales increase to near the mass of the weak bosons, the weak force becomes comparable to the other forces. All fermions can interact with the weak bosons (including neutrinos) and the W and Z bosons can also interact with itself. Leading order vertices of the weak force are plotted in Fig. 2.4.

The best example of weak interactions in everyday life is that of beta decay. In this form of radioactivity, a neutron (or proton), within a nucleus, decays to a proton

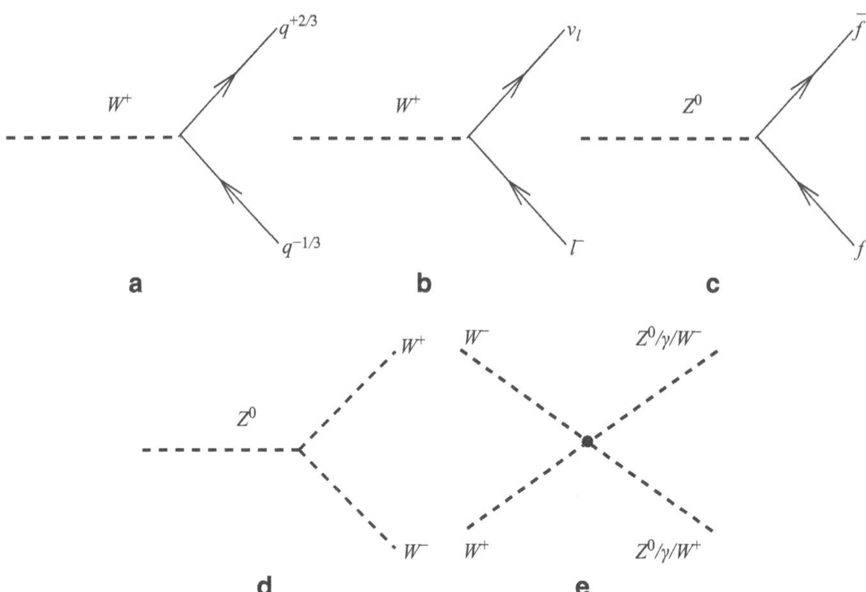

Fig. 2.4 Leading order weak interaction vertices used in Feynman diagrams: **a** $W^{\pm}$–up-type-quark–down-type-quark, **b** $W^{\pm}$–charged-lepton–neutrino, **c** Z^0–fermion–fermion, **d** Z^0–W^+–W^-, and **e** two $W^{\pm}$'s interacting with either two Z^0's, two γ's or two $W^{\pm}$'s

(neutron) and emits both an electron (positron) and an electron antineutrino (electron neutrino). The decay of muons is also mediated by the weak force, which explains its long lifetime.

2.1.4 The Higgs Mechanism

The theories mentioned above describe the fundamental forces of the SM. Since the SM was developed in the 1960s, nearly every prediction has been verified and every theorized particle has been found. However, there is one wrinkle to the SM: the masses of the weak bosons and the fundamental fermions.

In physics, it is generally understood that for each symmetry in nature, there is an associated conservation law. For example, symmetry in time implies a conservation of energy, symmetry in space implies a conservation of momentum, and symmetry in rotation implies a conservation of angular momentum. There exist other symmetries and other conservation laws that are a little more abstract. In QED, the conservation of charge is implied by the symmetry (or invariance) of applying local gauge transformations, which also accounts for the interaction between charged particles and the photon. Similarly, local gauge invariance also accounts for the conservation of colour charge in QCD, and describes the interaction of quarks with gluons.

The weak regime, alone, does not satisfy local gauge invariance. Unlike QED and QCD, which have massless bosons, the weak regime, with its massive bosons, becomes non-unitary within the framework of the SM. This symmetry, however, is restored with the introduction of a scalar field with non-zero vacuum expectation value. As the SM is perturbative in nature, the Lagrangian of the system is transformed to involve expansions about this vacuum expectation value. Although this mechanism does introduce massive gauge bosons, massless "ghost" particles (known as Goldstone particles) also appear. Fortunately, a proper gauge transformation can be chosen, such that the gauge field "eats" up these Goldstone particles. This apparent extra degree of freedom actually accounts for the additional polarization[4] of the massive gauge boson. For a detailed explanation of how a field acquires mass, see Appendix A.

This mechanism is known as the "Higgs mechanism" [9–14], named after the writer of one of the original papers, Peter Higgs. Although the explanation above (along with the example in Appendix A) is a simplified description of the method, the electroweak ($SU(2)_L \times U(1)$) gauge symmetry is restored with the addition of this new scalar field.

The Higgs field can be used to generate masses for fermions as well, using the same mechanism. The coupling of a quark or lepton field with the Higgs field (generally known as a Yukawa interaction), along with an appropriate gauge transformation, causes the fermions to acquire mass, just as the massive bosons did above.

The Higgs field, in turn, has its own boson (the Higgs boson), which couples to the massive particles of the SM, as plotted in Fig. 2.5, with the strength of the coupling directly proportional to the mass of the other particle. As the Higgs boson itself has mass, it also couples to itself.

2.1.5 *Interactions of the Standard Model*

As mentioned in each of the previous sections, all the particles of the SM interact with other particles in very specific ways. These interactions can be predicted using the fundamental theory of the relevant force. To calculate the predictions at leading order of a certain process of the SM, the most basic combination of vertices (using the ones pictured above) is used to build the given process. Predictions are calculated using the Feynman calculus, a set of rules based off the number and types of particles and vertices. It can be used to calculate particle lifetimes or reaction cross sections. Typically denoted as σ, cross section is a measure of the likelihood of an interaction to occur between two particles. The actual unit of σ is area, and was derived from the classical picture of point-like particles being fired at an area that includes a solid target. The point-like particles will either scatter or not, depending on whether it hits

[4] Massless particles have two polarizations, in the transverse plane. Massive particles, however, have an additional polarization, in the longitudinal direction.

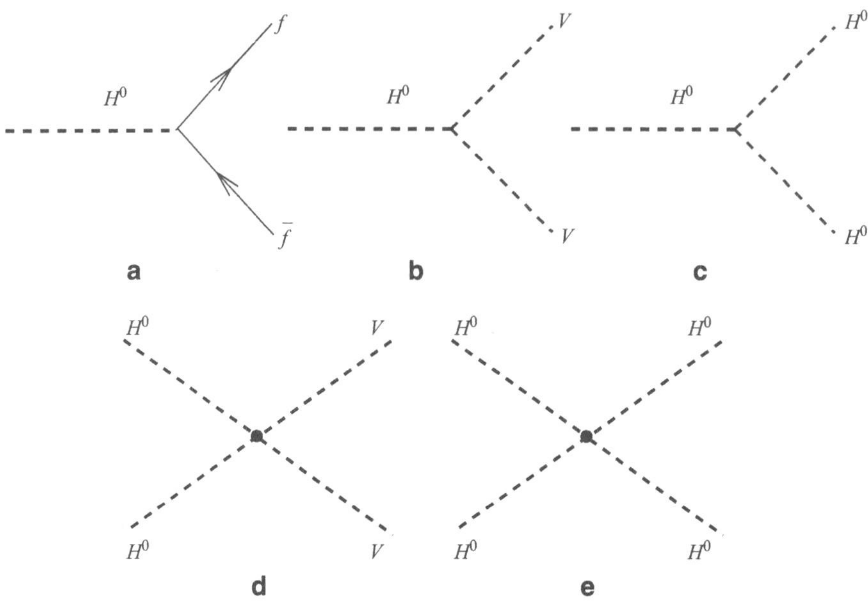

Fig. 2.5 Leading order vertices involving the SM Higgs boson to massive particles: **a** fermions, f, **b** vector bosons, V, **c** Higgs self-coupling, **d** di-Higgs to dibosons, and **e** four Higgs self coupling

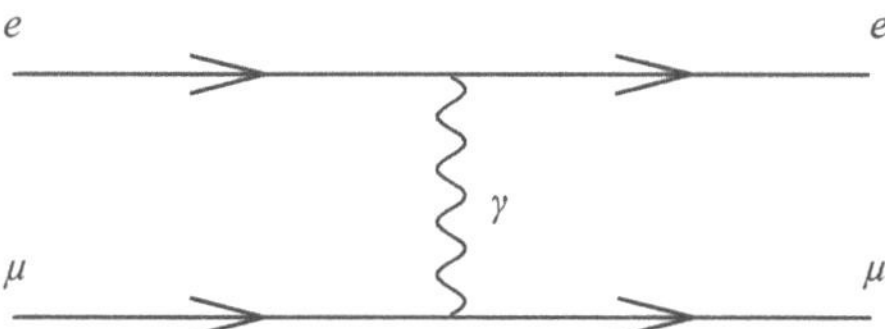

Fig. 2.6 Most basic Feynman diagram of electron-muon scattering

the solid target. The "interaction" probability is the ratio of cross sectional area of the solid target and the total area in which the targets are being fired. In particle physics, the cross sections are so small (on the order of the size of subatomic particles), that the more common unit to use is the "barn", where 1 b $= 10^{-24}$ cm^2.

The use of Feynman diagrams can be demonstrated using electron-muon scattering, pictured in Fig. 2.6. The plot shows the lowest order diagram of electron-muon scattering, as it is drawn using the smallest number of vertices. However, more complex diagrams exist that have the same initial and final particles, but with additional vertices. Examples of these higher order Feynman diagrams are drawn in Fig. 2.7. In theory, each of these additional diagrams contributes to the scattering calculation. However, each additional QED vertex introduces a factor of $\alpha \approx 1/137$, meaning they contribute much less to the overall scattering cross section calculation. Thus, to first-order, Fig. 2.6 alone can be used to predict electron-muon scattering. These

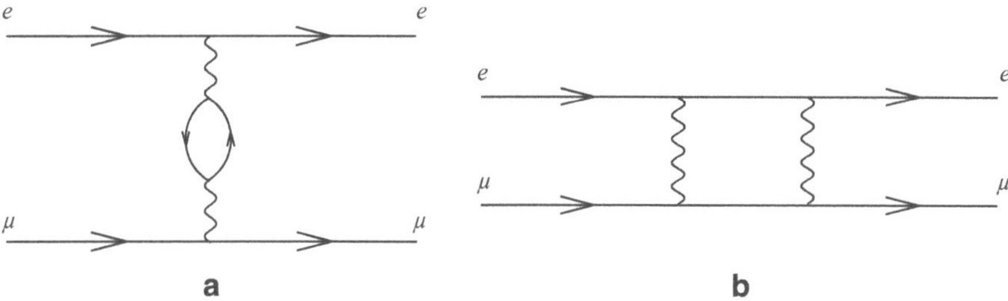

Fig. 2.7 Examples of higher order diagrams that contribute to the electron-muon scattering process

higher order diagrams are often omitted because they add much more complexity to the scattering calculation; however, approximate methods (e.g. perturbation theory) may be used to add these diagrams to the calculation, thus achieving a more precise prediction. The process can be further complicated by the addition of other final state particles (for example the production of an additional photon in electron-muon scattering). These are examples of next-to-leading order diagrams, and, though they marginally contribute to the final cross section, must be considered to derive more precise calculations.

2.2 Proton–Proton Collisions

Most of the particles of the SM have short lifetimes and cannot be found by themselves in nature. This makes the study of these particles rather difficult. In order to more thoroughly understand and study the particles of the SM, it is necessary to produce them artificially. This can be done in one of several ways; however, the most common way is by colliding stable energetic particles together that have accessible energies greater than the rest mass of the particle of interest. The available energy in a collision between two particles is often characterized using one of the Mandelstam variables, s [15]:

$$s \equiv (p_\mu^A + p_\mu^B)^2 = (E_A + E_B)^2 - (\mathbf{p_A} + \mathbf{p_B})^2 \tag{2.1}$$

where p_μ^A is the four-momentum[5] of particle A before collision, E_A is its energy, and $\mathbf{p_A}$ is its momentum vector. The available total energy is maximized in the centre-of-mass frame ($\mathbf{p_A} = -\mathbf{p_B}$), in which case the total centre-of-mass energy is:

$$E_{CM}^{TOT} = \sqrt{s} = E_A + E_B. \tag{2.2}$$

[5] Four-momentum is simply an extension of the classical three-dimensional momentum, $\mathbf{p} = (p_x, p_y, p_z)$, to also contain energy, $p_\mu \equiv (E, \mathbf{p}) = (E, p_x, p_y, p_z)$.

This energy can then be converted into a massive particle of mass up to $\sqrt{s}$. For example at LEP, in Geneva, Switzerland in the late 1980s, by colliding a beam of electrons of 45 GeV with a beam of positrons of same energy (amounting to $\sqrt{s} \approx 90$ GeV), physicists were able to effectively produce and study the Z^0 boson.

Electron-positron collisions may be the simplest to understand as they annihilate with each other. However, any collision of a pair of particles is also capable of producing more massive particles if their centre-of-mass energy, $\sqrt{s}$, is high enough. The LHC collides protons due to their relative ease to obtain, their ability to achieve high rates of collisions, and their minimal energy loss due to radiation when accelerated in a circle (ie. the LHC tunnel), compared to electrons. However, several additional issues do arise when colliding protons. Protons are not fundamental particles, but rather are made up of three valence quarks, that are embedded in a "sea" of quark-antiquark pairs, generated by the gluons exchanged between the quarks of the proton. Fortunately, due to asymptotic freedom, as the proton energy increases, the coupling between the partons within decreases, allowing the partons to basically move freely and independently of each other within the proton. The partons of the proton are thus better described using the variable x, the fraction of the total momentum of the proton. The momentum distribution functions of the parton are then described using Parton Distribution Functions (PDFs). The PDFs are basically a probability density to find a parton carrying a momentum fraction x, at a given energy scale, Q^2 [6].

Each parton type has its own PDF, though their general shape is higher at low x, and dropping down to 0 at high x. However, as Q^2 increases, the valence quarks are no longer dominant, with more quark-antiquark pairs created within the proton. Therefore, as Q^2 increases, the densities increase at low x, since the proton's momentum becomes more spread through the partons of the proton. This is seen clearly in Fig. 2.8, which show Next-to-Leading Order (NLO) PDFs used at the LHC, derived using data from several deep inelastic scattering experiments over the last 20 years. It is also interesting to note that the quarks and antiquarks typically carry about 50 % of the proton's momentum, with gluons carrying the other half. The gluon fraction also tends to increase with Q^2.

PDFs are very important in proton-proton collisions, since they are essential in predicting scattering processes. For example, calculating the cross section for a specific process, $a + b \to X$, in a proton-proton collision, is accomplished by evaluating the following:

$$\sigma_{ab \to X} = \sum_{i,j} \int dx_1 dx_2 f_i(x_1, Q^2) f_j(x_2, Q^2) \hat{\sigma}_{ij}(x_1 P_1, x_2 P_2, Q^2), \tag{2.3}$$

[6] Q^2 by definition is equal to $-t$, where t is another Mandelstam variable, defined as $t \equiv (p^A_{\mu,i} + p^A_{\mu,f})^2 = (E_{A,i} + E_{A,f})^2 - (\mathbf{p}_{\mathbf{A,i}} + \mathbf{p}_{\mathbf{A,f}})^2$, where the variables are now the initial and final energies and momenta of the same article. This is essentially a measure of momentum transfer in a collision, where a "soft", glancing collision leads to a small Q^2, whereas a "hard", more direct collision leads to a larger Q^2.

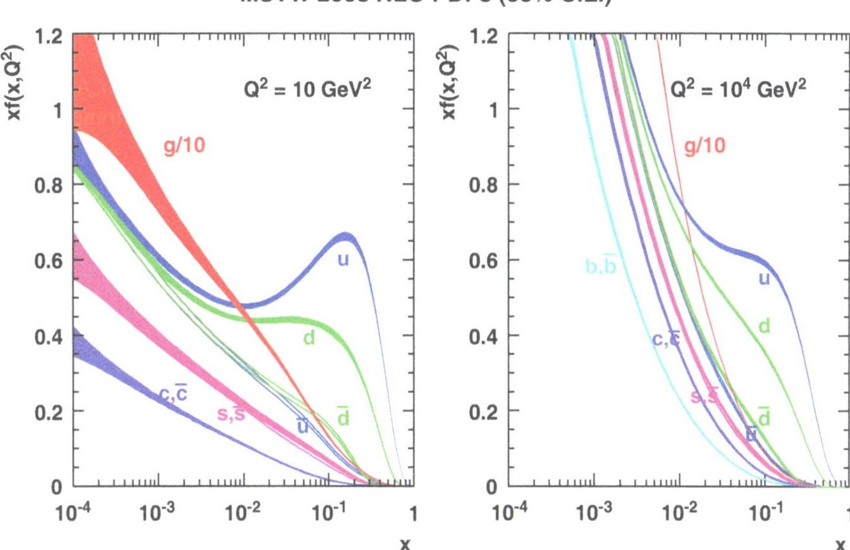

Fig. 2.8 Parton distribution functions of the quarks and gluons within a proton at a lower energy scale ($Q^2 = 10$ GeV^2, *left*) and at a higher energy scale ($Q^2 = 10^4$ GeV^2, *right*) at NLO. The vertical axis is $x \cdot f(x, Q^2)$, the parton fraction, x, times the distribution function. The gluon distribution is scaled by a factor of 1/10 for display purposes. Taken from [16]

where i, j are the possible initial state partons (gluon, up quark, etc.), f_i and f_j are the PDFs for the given parton, P_1 and P_2 are are the initial momenta of the protons, and $\hat{\sigma}_{ij}$ is the cross section for the specific process, derived as a function of initial state momenta and energy scale, Q^2. The cross section is typically derived using the Feynman calculus, as described in Sect. 2.1.5. This method is only valid for hard scattering processes (those at high energy scales, Q^2, or at small length scales), where a perturbative approach can be used to calculate $\hat{\sigma}$. Soft processes (low energy scales or long length scales) are dominated by non-perturbative QCD effects, which are much less well understood [17].

2.2.1 Coordinate System in Hadron Collisions

The coordinate system used in most hadron collisions is a right-handed cylindrical system, where the origin lies at the interaction point, with the z-axis running along the beam line. The plane that is transverse to the beam, or x–y, is very important in hadron collisions. As seen in Eq. 2.3, the cross section for a process is calculated by integrating over all *possible* values of x, as plotted, for example, in Figure 2.8. For any given collision at the LHC, however, the value of x for either colliding parton is unknown. Thus, it is not known, *a priori*, whether the collision occurred at centre-of-mass (ie. with momenta summing to zero) or not. As the protons collide head-on along the z-axis, the initial momenta of the interacting partons in the transverse plane, x–y, are negligible. Thus, for any hard process, the vector sum of the transverse momenta of final state particles must add to zero. Quantities such as transverse momentum:

$$p_T \equiv \sqrt{p_x^2 + p_y^2} \tag{2.4}$$

and transverse energy:

$$E_T \equiv \sqrt{m^2 + p_T^2} \tag{2.5}$$

are often used. It also becomes convenient to use an adapted four-momentum notation:

$$p_\mu = (E, p_x, p_y, p_z) \tag{2.6}$$

$$= (m_T \cosh y, p_T \cos\phi, p_T \sin\phi, m_T \sinh y) \tag{2.7}$$

where ϕ is the azimuthal angle, $m_T = \sqrt{E^2 - p_T^2}$ is the transverse mass, and y is the rapidity:

$$y \equiv \frac{1}{2} \ln\left(\frac{E + p_z}{E - p_z}\right). \tag{2.8}$$

The advantage of using these quantities are that p_T, ϕ and differences in y are invariant under longitudinal boosts.

For particles travelling close to the speed of light (or when $|\mathbf{p}| \gg m$), rapidity can be approximated to pseudorapidity, η:

$$\eta \equiv \frac{1}{2} \ln\left(\frac{|\mathbf{p}| + p_z}{|\mathbf{p}| - p_z}\right) = -\ln\left[\tan\left(\frac{\theta}{2}\right)\right], \tag{2.9}$$

where θ is the polar angle in the cylindrical coordinate system. To help visualize this quantity, $\eta = 0$ corresponds to a vector pointing in the transverse plane, whereas $\eta = +(-)\infty$ corresponds to vectors in the positive (negative) beam axis. Another advantage of this quantity is that for minimum bias events [7], the particle multiplicity per unit of rapidity is approximately constant.

A useful quantity to measure angular separation in the detector is ΔR:

$$\Delta R = \sqrt{(\eta_1 - \eta_2)^2 + (\phi_1 - \phi_2)^2}. \tag{2.10}$$

This, along with the other quantities above, are used throughout this thesis.

[7] Minimum bias events are ones that would be collected with a totally inclusive trigger, and would include both diffractive and non-diffractive events. Diffractive events occur when the protons are not, or just barely, broken up; non-diffractive events occur when the protons are broken up and hit the detector. Diffractive events are experimentally difficult to measure, thus most minimum bias events are inelastic non-diffractive collisions.

Chapter 3
Status of Higgs Boson Search and Discovery

The Higgs mechanism, described in Sect. 2.1.4, was first theorized nearly 50 years ago to explain why certain particles, notably the weak bosons, had mass. Although this mechanism does successfully explain the mass of the W and Z bosons[1], it does not predict the mass of the particle associated to it: the Higgs boson. This has been an open question for particle physicists ever since. Some methods exist to indirectly estimate the mass of the Higgs; however, due to the very weak dependence of the Higgs mass on other observables, many experiments conducted direct searches for this elusive particle [18].

3.1 SM Higgs Predictions

Though the SM does not predict the mass of the SM Higgs boson[2], it does predict how the Higgs can be produced and how it can decay. The production modes of the Higgs boson depend on the type of collisions used to produce them. At e^+e^- colliders (such as LEP), the primary method to produce the Higgs is via "Higgsstrahlung", in which the e^+e^- collision produces a Z^0, which radiates a Higgs boson (ie. $e^+e^- \to H^0Z^0$). At hadron colliders (such as the Tevraton and the LHC), the mechanism with the highest probability to produce a SM Higgs is gluon-gluon fusion, plotted in Fig. 3.1a, where a Higgs is produced from a top-quark loop created by a pair of gluons. The second most abundant is VBF, plotted in Fig. 3.1b, where colliding quarks exchange a pair of weak bosons that inverse pair decay into a Higgs. The next most abundant is Higgstrahlung, or the associated production with a weak boson, plotted in Fig. 3.1c, where a produced weak boson (either $W^\pm$ or Z^0) radiates a Higgs. Finally, the last one often considered is the associated production with a top-antitiop quark pair, plotted in Fig. 3.1d. The cross sections for each of these production mechanisms

[1] Given the additional input parameter $\sin^2\theta_W$.

[2] The "SM Higgs boson" is a term used throughout this thesis and any paper related to the Higgs searches. It is defined as the Higgs boson that couples to all other particles exactly as predicted by the SM, given the mass of the Higgs.

E. Ouellette, *Search for the Higgs Boson in the Vector Boson Fusion Channel at the ATLAS Detector*, DOI 10.1007/978-3-319-13599-1_3

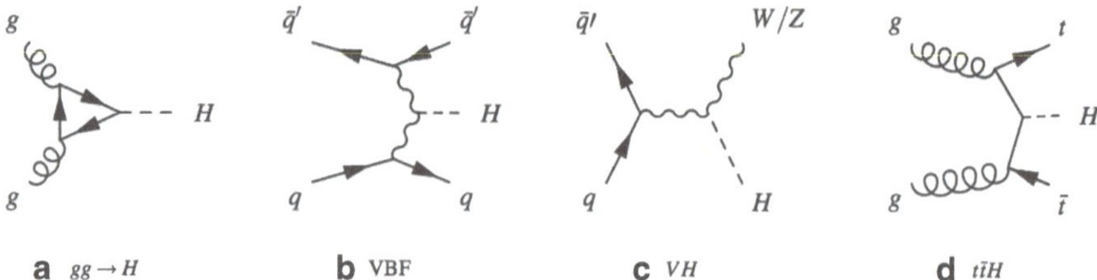

Fig. 3.1 Feynman diagrams of the most abundant Higgs production mechanisms at hadron colliders: **a** gluon-gluon fusion, **b** vector boson fusion, **c** associated production with a weak boson and **d** in association with a top-antitop pair. Taken from [19]

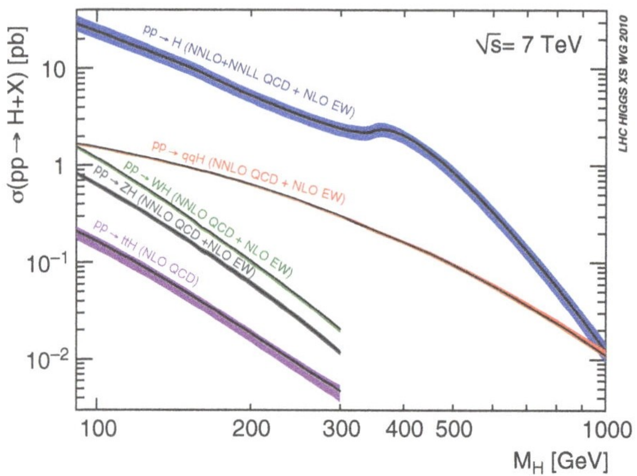

Fig. 3.2 Higgs production cross sections as a function of Higgs mass at $\sqrt{s} = 7$ TeV pp collisions. $pp \to H$ corresponds to gluon-gluon fusion, $pp \to qqH$ to VBF, $pp \to VH$ (where $V = W$ or Z) to the associated production with a weak boson, and $pp \to ttH$ to the associated production with a top-antitop quark pair. Calculations were not performed for $m_H > 300$ GeV for the bottom three processes, as their cross sections were deemed too low. Taken from [20]

are dependent on Higgs mass, with cross section decreasing with mass, as shown in Fig. 3.2.

The Higgs boson couples to any massive particle, though the strength of the coupling is directly proportional to mass. For this reason, the Higgs (of any mass) decays mostly to the heaviest kinematically attainable particle pair. Thus, as shown in Fig. 3.3, lower Higgs masses ($<$ 135 GeV) prefers $H^0 \to b\bar{b}$ decays, whereas heavier Higgs' prefer $H^0 \to W^+W^-$.

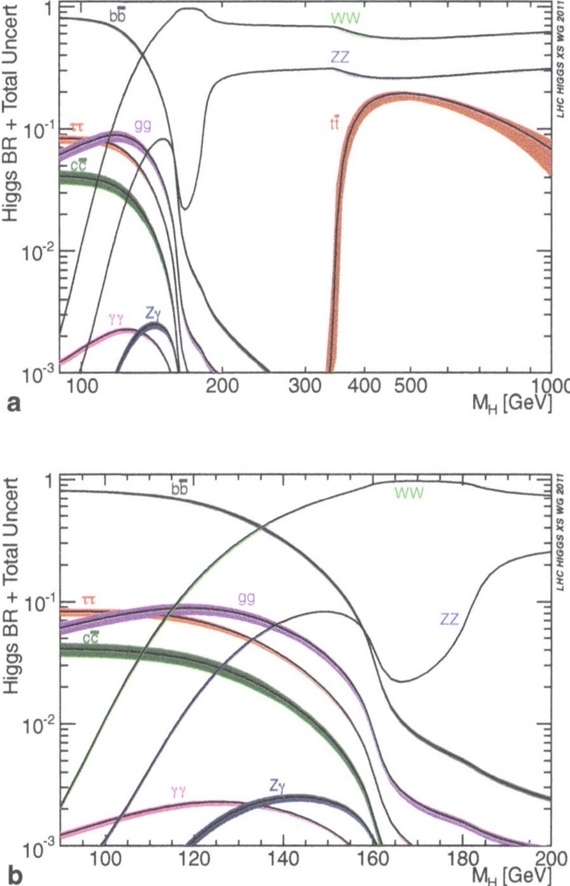

Fig. 3.3 Higgs branching ratios as a function of Higgs mass. **a** shows the mass range of $100 < m_H < 1000$ GeV, while **b** shows a zoomed in region of the previous plot. Taken from [20]

3.2 Searches at the Large Electron-Positron Collider

Early limits for the Higgs boson mass were derived from neutron-nucleus scattering [21] and other nuclear physics experiments; however, the first direct searches were conducted at the Large Electron-Positron (LEP) collider at CERN. During its 11 year run, LEP was able to set a lower limit on the mass of the Higgs of $m_H > 114.4$ GeV [22] at the 95 % Confidence Level (CL) (see Appendix C for more details on CL limits). This was accomplished by searching for Higgsstrahlung produced Higgs' (ie. $e^- e^+ \rightarrow H^0 Z^0$), and combining results from various Higgs and Z^0 decay modes, from each of the four LEP experiments. The primary Higgs decay channel was $H^0 \rightarrow b\bar{b}$, since this was the dominant decay channel for the Higgs mass range for which the experiment was sensitive ($m_H < 115$ GeV).

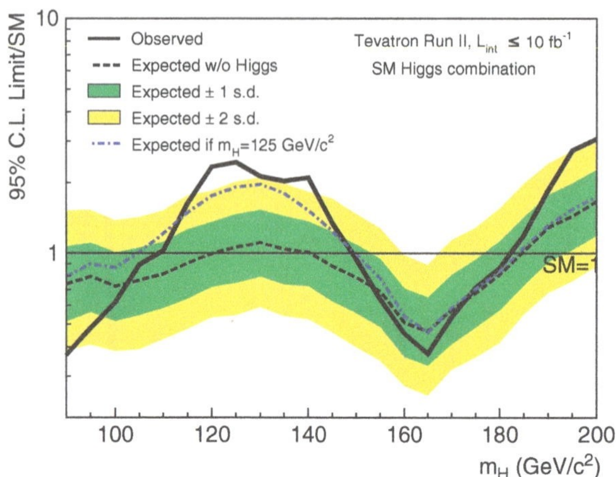

Fig. 3.4 Observed and expected 95 % CL limits of the SM Higgs cross section using Tevatron data, as a function of m_H. The *black dashed line* corresponds to the expected upper limit on the Higgs cross section, derived from signal and background samples (along with the $\pm 1\sigma$ and $\pm 2\sigma$ bands, in *green* and *yellow*, respectively). The *solid black line* is the observed upper limit on the SM cross section, where for any mass for which this lines goes below 1, the SM Higgs is excluded. The *blue short-dashed line* shows the median expected limits assuming a 125 GeV Higgs is present [23]

3.3 Searches at the Tevatron Collider

In parallel to LEP, Higgs searches were also conducted at the Tevatron collider at the Fermi National Accelerator Laboratory in Batavia, Illinois USA. The Tevatron collided proton-antiproton beams with a maximum centre-of-mass energy of $\sqrt{s} =$ 1.96 TeV at one of two detectors: CDF and DØ. The background at hadron colliders is much greater than at e^+e^- colliders. However, the higher centre-of-mass energies of the Tevatron allowed for searches of higher Higgs masses than LEP allowed. Using all the Higgs production modes in Fig. 3.1, along with the Higgs decay channels of $H^0 \rightarrow b\bar{b}$, $H^0 \rightarrow W^+W^-$, $H^0 \rightarrow \tau^+\tau^-$, and $H^0 \rightarrow \gamma\gamma$, the Tevatron was able to exclude the SM Higgs boson in the mass range of $90 < m_H < 109$ GeV and $149 < m_H < 182$ GeV at the 95 % CL, using nearly 10 fb^{-1} of data [23]. The final results show an excess of events in the mass range of $115 < m_H < 140$ GeV. The observed local significance at $m_H = 125$ GeV corresponds to a three standard deviation (3σ) deviation from a background-only hypothesis, which is consistent with the mass of the Higgs observed at the LHC. Results are plotted in Fig. 3.4.

The data samples used to obtain these results were collected during Run II, which ran from 2001 to 2011. Tevatron operations ceased in late 2011 due to the LHC achieving luminosity of almost ten times that of the Tevatron, and at energies already 3.6 times greater.

3.4 Searches at the Large Hadron Collider

The search for the Higgs boson has been a cornerstone of the physics programs at both A Toroidal LHC ApparatuS (ATLAS) and the Compact Muon Solenoid (CMS) experiments of the LHC at the European Organization for Nuclear Research (CERN).

3.4.1 Higgs Discovery at ATLAS

Results from Higgs searches at ATLAS have been released periodically since data taking began at the LHC in 2010. Significant progress was made after the $\sqrt{s} = 7$ TeV 2011 data run, where roughly 5 fb^{-1} of data was collected. Several Higgs masses were excluded at the time, with hints of a possible Higgs around 126 GeV. It was finally discovered in 2012 using up to 22 fb^{-1} of $\sqrt{s} = 8$ TeV data collected. Several different channels were used in the discovery; however, the so-called 'Golden Channels' of $H^0 \rightarrow \gamma\gamma$ and $H^0 \rightarrow ZZ^{(*)}$ had the largest impact on the discovery.

$H^0 \rightarrow \gamma\gamma$ Channel

The $H^0 \rightarrow \gamma\gamma$ channel used events with at least two photons that satisfy a range of optimized selection criteria [24, 25]. Using 4.8 fb^{-1} of $\sqrt{s} = 7$ TeV data and 20.7 fb^{-1} of $\sqrt{s} = 8$ TeV data, a clear excess was seen around $m_H = 126.8$ GeV, as plotted in Fig. 3.5. This excess was quantified by calculating a local p_0 value, which measures the probability of the background to fluctuate beyond the observation in data at a particular value of m_H. The result of the p_0 calculation shows a minimum observed (expected) significance of 7.4(4.1)σ at $m_H = 126.5$ GeV. When considering the chance of fluctuations at any given mass value (the so-called "look-elsewhere effect"), this significance drops to 6.1(2.9)σ.

A mass measurement calculation was also derived using a profile likelihood ratio test statistic, giving a best-fit measurement of $m_H = 126.8 \pm 0.2(\text{syst}) \pm 0.7(\text{stat})$ GeV. Using this mass, a signal strength calculation was also performed. The signal strength parameter, μ, is a measure of how strong a possible signal may be when conducting a particle search, with $\mu = 0$ corresponding to a background-only hypothesis, and $\mu = 1$ being the nominal signal hypothesis (see Appendix C for a more thorough description of the statistical analysis). The best-fit value for μ was $1.65 \pm 0.24(\text{stat})^{+0.25}_{-0.18}(\text{syst})$, which corresponds to a 2.3σ deviation from the SM prediction of signal-plus-background.

$H^0 \rightarrow ZZ^{(*)}$ Channel

The $H^0 \rightarrow ZZ^{(*)} \rightarrow 4l$ channel has always been considered the 'Golden Channel', due to its very clean four lepton final state, with very low background, and good mass resolution. Four different final states were considered for this analysis: $\mu^+\mu^-\mu^+\mu^-$

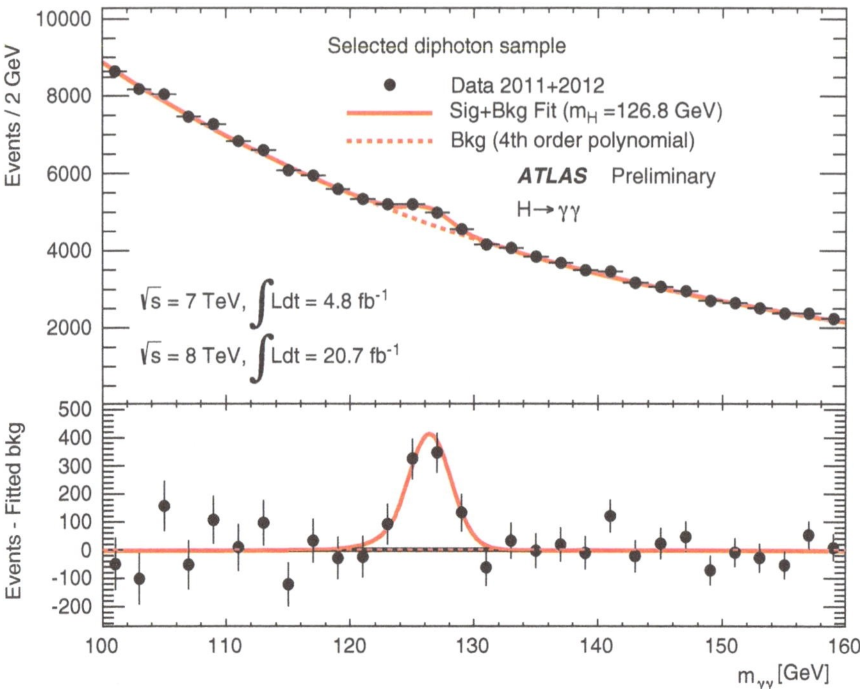

Fig. 3.5 Invariant mass distribution of $m_{\gamma\gamma}$ for combined 2011 and 2012 datasets for $H^0 \to \gamma\gamma$ search channel at ATLAS. A fit to the data shows the sum of a signal component (with $m_H =$ 126.8 GeV) and the background. A clear excess is shown around the considered signal mass point. Taken from [25]

(4μ), $\mu^+\mu^-e^+e^-$ $(2\mu 2e)$, $e^+e^-\mu^+\mu^-$ $(2e2\mu)$ and $e^+e^-e^+e^-$ $(4e)$[3]. Using 4.6 fb^{-1} of $\sqrt{s} = 7$ TeV data, and 20.7 fb^{-1} of $\sqrt{s} = 8$ TeV data [26, 27], a clear excess was seen around 125 GeV, as plotted in Fig. 3.6. A 95 % CL upper limit on the SM Higgs cross section was calculated over the entire considered mass range, and $m_H > 130$ GeV was excluded. This is plotted for the lower Higgs mass range in Fig. 3.7, though with a clear excess in the range of 115 to 130 GeV. Similar to $H^0 \to \gamma\gamma$, a local p_0 value calculation was conducted, giving a minimum observed (expected) local p_0 value of $6.6(4.4)\sigma$ at $m_H = 124.3$ GeV.

Mass measurement and signal strength calculations were also derived using a profile likelihood ratio test statistic. The best-fit measurement of the Higgs mass was:

$$m_H = 124.3^{+0.6}_{-0.5}(\text{stat})^{+0.5}_{-0.3}(\text{syst})\ \text{GeV}.$$

[3] Note that $2\mu 2e$ and $2e2\mu$ differ in the lepton flavour pair with invariant mass closest to the Z^0 mass.

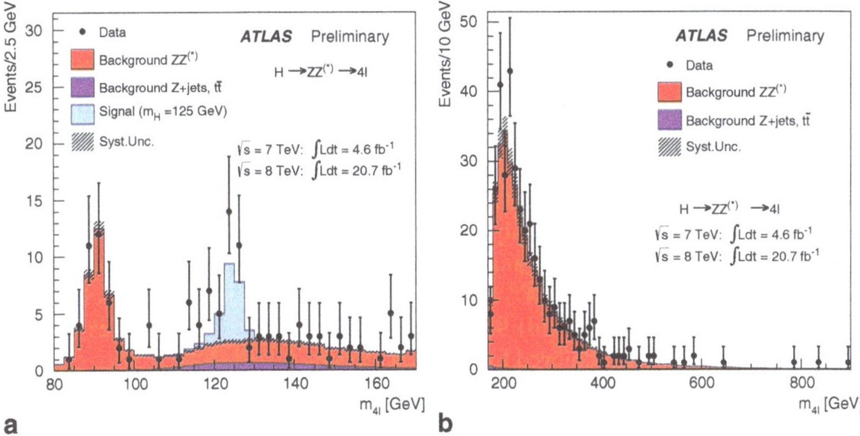

Fig. 3.6 Invariant mass distributions of m_{4l} for combined 2011 and 2012 datasets for $H^0 \rightarrow ZZ^{(*)} \rightarrow 4l$ search channel at ATLAS in the mass range (**a**) 80–170 GeV and (**b**) 170–900 GeV. A $m_H = 125$ GeV signal expectation is also shown. Taken from [27]

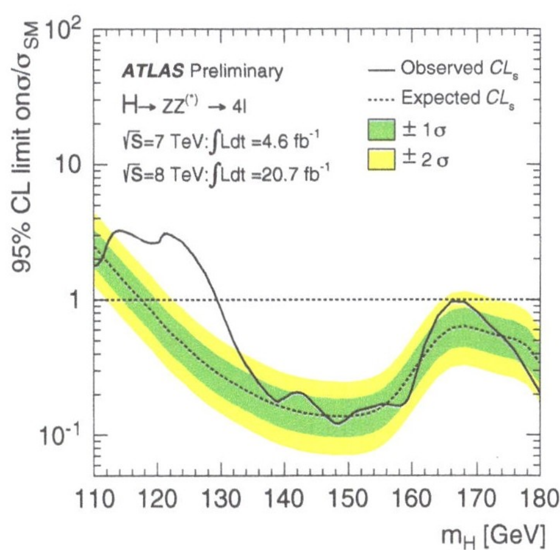

Fig. 3.7 Expected (*dashed*) and observed (*solid line*) 95 % CL upper limit on SM Higgs cross section as a function of m_H using combined 2011 and 2012 data for the $H^0 \rightarrow ZZ^{(*)} \rightarrow 4l$ search channel at ATLAS. Taken from [27]

Using this mass in the signal strength calculation, the best-fit μ value is $1.7^{+0.5}_{-0.4}$.

Other Channels

Several other channels were also considered. After the excess was discovered around 125 GeV, there was a big push to look at the other dominant decay channels in that region to further help with the particle discovery.

An $H^0 \rightarrow WW^{(*)} \rightarrow e\nu\mu\nu$ analysis was developed for $\sqrt{s} = 8$ TeV data [28]. These events are difficult to identify due to the missing energy from the two neutrinos. Since the invariant mass cannot be directly measured, an adapted transverse mass variable is used instead. Results from the analysis show an excess of events for $m_H \lesssim 150$ GeV, with a 2.6σ excess at 125 GeV.

The $H^0 \rightarrow \tau^+\tau^-$ analysis, on both $\sqrt{s} = 7$ TeV and $\sqrt{s} = 8$ TeV data [29], utilizes two different τ decay types: hadronic and leptonic. The analysis used several different categories, which allows for ggF, VBF and VH production mechanisms to each be considered. The observed (expected) 95 % CL upper limit on the SM cross section was 1.9 (1.2) times the SM prediction for $m_H = 125$ GeV.

The $H^0 \rightarrow b\bar{b}$ channel used only the VH production mechanism for its combined $\sqrt{s} = 7$ TeV and $\sqrt{s} = 8$ TeV analysis [30]. Three different vector boson decays were considered: $Z^0 \rightarrow \nu\nu$, $W^{\pm} \rightarrow l^{\pm}\nu$ and $Z^0 \rightarrow l^+l^-$. Using m_{bb} as a test statistic, an observed (expected) 95 % CL upper limit of 1.8 (1.9) times the SM Higgs cross section was achieved for $m_H = 125$ GeV.

Combined Results

Results from all channels were combined using a thorough statistical procedure (to account for correlated systematic uncertainties across different channels). The combined mass measurement used only the $H^0 \rightarrow \gamma\gamma$ and $H^0 \rightarrow ZZ^{(*)}$ channels, giving $m_H = 125.2 \pm 0.3(\text{stat}) \pm 0.6(\text{syst})$ GeV [31]. A combined signal strength calculation was also conducted, using all considered channels, leading to an average value of $\mu = 1.35 \pm 0.19(\text{stat}) \pm 0.15(\text{syst})$ computed for $m_H = 125$ GeV. A comparison of the signal strength results from each of the considered channels is plotted in Fig. 3.8. A compatibility test between the signal strength of each channel and the SM expectation of unity gave a probability of 13 %. The properties of the discovered particle are consistent with a SM Higgs boson.

3.4.2 *Higgs Discovery at CMS*

CMS, the other multi-purpose detector at the LHC, has also been searching for the Higgs since data taking began in 2010. Although the search channels used by CMS are similar to those used by ATLAS, the methods used are different, as both experiments run independently of each other. The results from CMS are consistent with those from ATLAS, with CMS also discovering a Higgs boson around $m_H \approx 125$ GeV [32].

The $H^0 \rightarrow ZZ^{(*)} \rightarrow 4l$ channel had a very large impact on the Higgs discovery at CMS. Its analysis differed from ATLAS by including the Z^0 decay to $\tau^+\tau^-$. Using 5.1 fb^{-1} of $\sqrt{s} = 7$ TeV data, and 19.6 fb^{-1} of $\sqrt{s} = 8$ TeV data [33], a clear excess is seen, as plotted in Fig. 3.9, in the combined four-lepton invariant mass distribution. This corresponds to a 6.7σ deviation above background expectation, with a best-fit mass of $m_H = 125.8 \pm 0.5(\text{stat.}) \pm 0.2(\text{syst.})$ and a best-fit signal strength of $\mu = 0.91^{+0.30}_{-0.24}$.

The $H^0 \rightarrow \gamma\gamma$ channel had a smaller impact on the CMS Higgs discovery, compared to ATLAS. Using 5.1 fb^{-1} of $\sqrt{s} = 7$ TeV data, and 19.6 fb^{-1} of $\sqrt{s} =$ 8 TeV data [34], both cut-based and multivariate technique-based analyses were

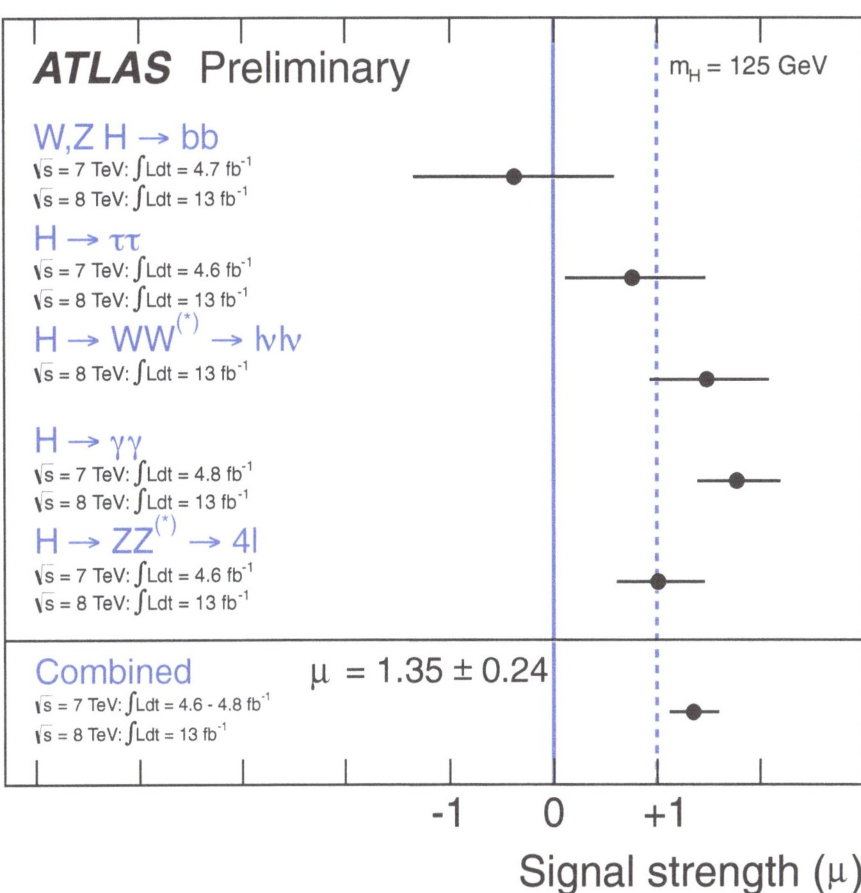

Fig. 3.8 Comparison of signal strength, μ, from each individual channel and a combined result for $m_H = 125$ GeV at ATLAS. Taken from [31]

developed to search for the Higgs in diphoton events. A clear excess was seen around 125 GeV, as plotted in Fig. 3.10, and corresponds to a local significance of 3.2σ (for the multivariate analysis). A best-fit signal strength calculation was conducted, giving a value of $\mu = 0.78 \pm 0.27$ at $m_H = 125$ GeV. The best-fit mass measurement was $m_H = 125.4 \pm 0.5(\text{stat.}) \pm 0.6(\text{syst.})$ GeV.

Other search channels were used in the Higgs discovery, such as $H^0 \to W^+W^-$, $H \to \tau^+\tau^-$ and $H^0 \to b\bar{b}$, though their impact was smaller compared to $H^0 \to ZZ^{(*)} \to 4l$. Observed excesses around $m_H = 125$ GeV in each of these channels ranged from 2.8 to 3.9σ. A combined best-fit mass of the Higgs was measured to be $m_H = 125.7 \pm 0.3(\text{stat.}) \pm 0.3(\text{syst.})$ GeV, and a combined best-fit signal strength for that mass was measured to be $\mu = 0.80 \pm 0.14$ [35], with the contributions from each channel plotted in Fig. 3.11.

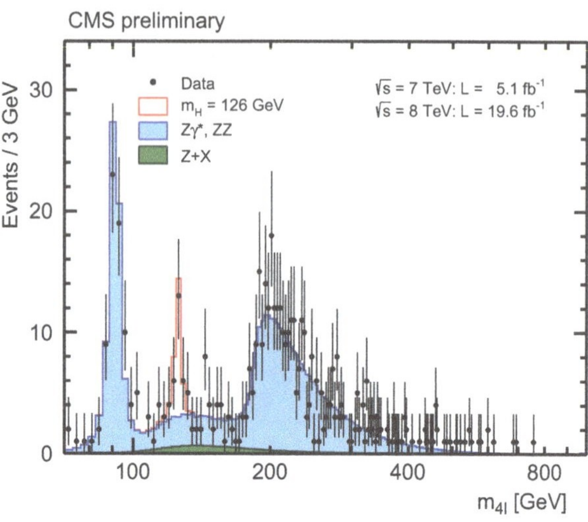

Fig. 3.9 Invariant mass plot of m_{4l} in $H^0 \to ZZ^{(*)} \to 4l$ channel at CMS. Points represent data, shaded histograms represent expected background and unshaded histogram represents the signal expectation for a 126 GeV Higgs boson. Taken from [33]

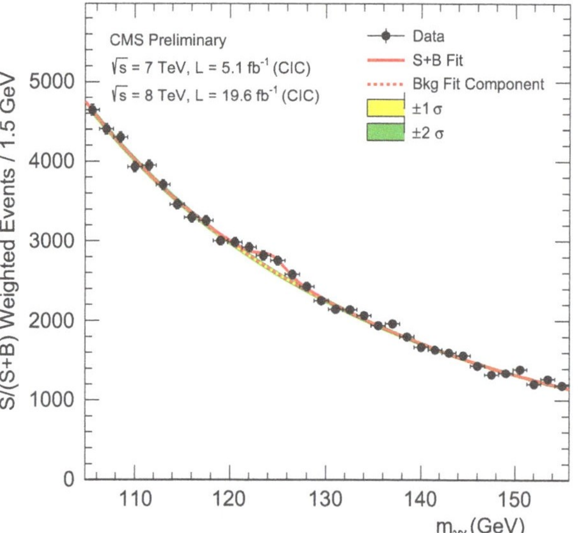

Fig. 3.10 Invariant mass plot of diphoton events in the $H^0 \to \gamma\gamma$ channel at CMS. Data events, represented at points, are weighted by the ratio of signal to signal-plus-background. Best-fits to background and signal-plus-background are plotted with $\pm 1\sigma$ and $\pm 2\sigma$ bands on the background only fit. Taken from [34]

3.5 Further Improvements and Analysis Motivation

The discovery of the Higgs boson at both ATLAS and CMS was a major success for the LHC and CERN. The next steps for both experiments are to refine their results, and to further determine the properties of this new particle. Recent results suggest the discovered boson is compatible with a spin-0, *CP*-even particle [36, 37]. However, more work is needed to confirm the discovery of the Higgs in the fermionic decay channels. So far, the Higgs discovery has not been confirmed in $H^0 \to b\bar{b}$ and

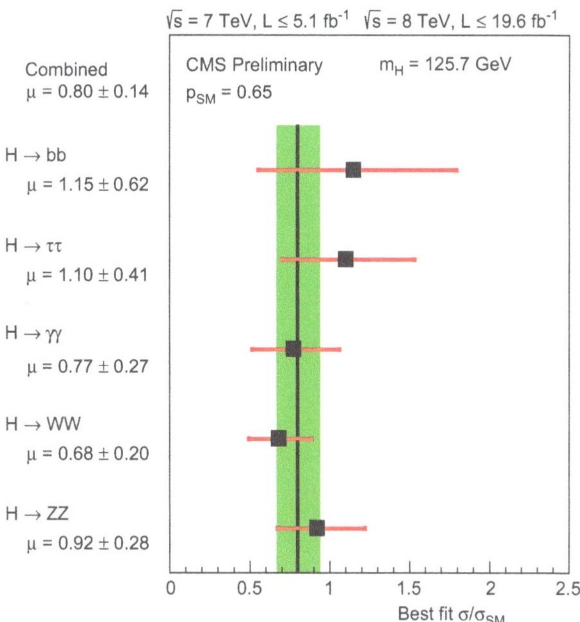

Fig. 3.11 Signal strength values for each Higgs decay channel used in the Higgs discovery at CMS. Taken from [35]

$H^0 \to \tau^+\tau^-$, the two most dominate fermionic decay channels. Slight excesses have been observed in each channel; however, as the error bars in their signal strength measurements in Figs. 3.8 and 3.11 suggest, nothing conclusive can be said just yet. The coupling of the Higgs to both massive bosons and fermions is predicted; therefore, the Higgs discovery should be confirmed in each of these channels for the SM prediction to be fully validated.

Most channels used for the Higgs discovery primarily utilized the gluon-gluon fusion production mode, as it is the dominate mechanism to produce a SM Higgs. Vector Boson Fusion (VBF), however, can be used to determine the *CP*-nature of the discovered Higgs, due to observables that are very sensitive to the Higgs boson properties [38].

These are two primary reasons to develop an analysis for the VBF $H^0 \to b\bar{b}$ channel. Further confirming the Higgs discovery in this channel would (a) be proof of fermionic decay of the Higgs, and (b) help in determining the properties of the Higgs. It would also contribute to the larger effort of quantifying the coupling of the Higgs to *b*-quarks.

Chapter 4
The Experiment

As mentioned in Chap. 3, the experiments of the Large Hadron Collider (LHC) have successfully discovered the Higgs boson. These state-of-the-art facilities have been built to find the Higgs in a wide variety of production and decay modes. Fortunately, it is well-suited to find the Higgs in the VBF $H^0 \rightarrow b\bar{b}$ channel.

4.1 The Large Hadron Collider

The LHC [39] is the centrepiece of the CERN laboratory in Geneva Switzerland. It is a 27 km circular accelerator, buried 50 to 175 m underground, beneath the Franco-Swiss border, capable of accelerating both protons and heavy ions. Separate beam pipes allow for two beams to circulate in opposite directions, which, at its design peak, can reach pp collisions at a centre-of-mass energy of $\sqrt{s} = 14$ TeV.

The LHC, pictured in Fig. 4.1, is a two-ring synchrotron particle accelerator. Its 1232 dipole magnets bend the beams into its circular orbit, whereas its 392 quadrupole magnets focus the beams. The dipoles' magnetic field required to bend the proton path (at the design energy) is 8.3 T. The resulting current and material requirements are beyond the capabilities of traditional electromagnets, thus, NbTi superconducting magnets that operate at 1.9 K are used instead. This is accomplished by using an elaborate cryogenics system, with liquid helium cooling the magnets to their operating temperature.

The LHC is the largest of the accelerators at CERN; however, a succession of small to large accelerators are used to accelerate the protons to the energy needed for injection into the LHC. Protons are first accelerated through LINAC 2, a linear accelerator, to reach an energy of 50 MeV. They are then injected into the Proton Synchrotron Booster (PS Booster), where their energies are increased to 1.4 GeV. Next comes the Proton Synchrotron (PS), where protons reach 26 GeV, before being injected into the Super Proton Synchrotron (SPS). Protons reach an energy of 450 GeV in the SPS, which is the injection energy of the LHC.

E. Ouellette, *Search for the Higgs Boson in the Vector Boson Fusion Channel at the ATLAS Detector*, DOI 10.1007/978-3-319-13599-1_4

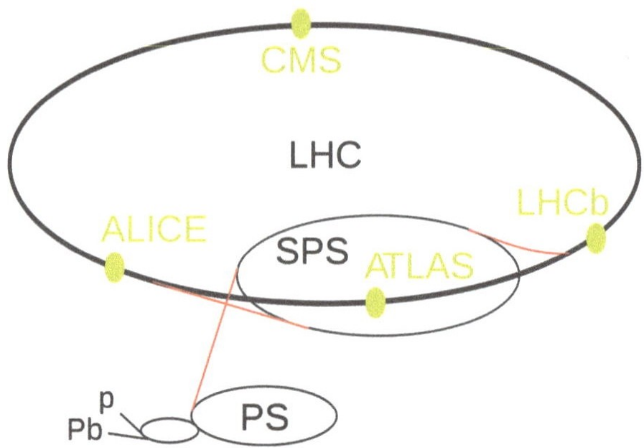

Fig. 4.1 Schematic view of the LHC complex at CERN in Geneva Switzerland. Taken from [40]

The protons are accelerated in the LHC in bunches, as they tend to group together during the acceleration process. There is an upper limit of 2808 proton bunches allowed in the LHC, which corresponds to a bunch crossing time of 25 ns.

Four interaction points exist around the LHC ring where the beams can collide. Major detectors have been placed at each of these points to identify and measure the particles created in these collisions. The detectors are: ATLAS and CMS, which are two very large multipurpose detectors, and ALICE and LHCb, smaller detectors specializing in heavy-ion collisions and b-meson physics, respectively.

4.1.1 LHC Luminosity

The rate at which events occur at the LHC is dependent on its luminosity. The instantaneous luminosity, $\mathcal{L}$, is the measure of the number of particles per unit area per unit time, with typical units $\text{cm}^{-2}\text{s}^{-1}$. This value is a function of several beam parameters, such as the number of protons per bunch, the number of bunches, revolution frequency, as well as many other beam specific characteristics. The LHC is designed to reach a peak luminosity of $10^{34}\ \text{cm}^{-2}\text{s}^{-1}$.

The number of events per second, dN/dt, for a given process is given by:

$$\frac{dN}{dt} = \sigma\mathcal{L} \tag{4.1}$$

where σ is the cross section. Over a long period of time, the number of events for the given process is simply the integral of the previous equation:

$$N = \sigma \int \mathcal{L}dt. \tag{4.2}$$

The cross section can be factored out of the integral as it is independent of time. The integral, $\int \mathcal{L}dt$, is referred to as the integrated luminosity, with units of cm^{-2}, though the use of pb^{-1} or fb^{-1} (inverse pico- or femto-barns) is more common, where $1\ \text{b} = 10^{-24}\ \text{cm}^2$.

4.2 The ATLAS Dectector

The A Toroidal LHC ApparatuS (ATLAS) detector [41] was constructed to measure and identify a wide variety of particles, which gives physicists the ability to observe several different types of processes. To accomplish this, ATLAS is nearly hermetic, covering almost the entire 4π area around the interaction point. ATLAS was also designed to be modular in nature, with each system accomplishing specific tasks independent of all other systems. ATLAS is typically divided into three major subsystems: the inner detector, the calorimeters and the muon detectors. The detector is pictured in Fig. 4.2.

4.2.1 The Inner Detector

The ATLAS inner detector is the system closest to the interaction point, measures the momentum of charged particles and measures primary and secondary vertices using charged tracks. This is done by immersing the inner detector in a 2 T magnetic field provided by a superconducting solenoid magnet (enveloping the inner detector), thus bending the path of charged particles as they pass through the detector. Tracks are reconstructed using hits in the multiple layers of the inner detector. Being located closest to the interaction point, the inner detector is subjected to the highest amount of radiation. It must also have very fast electronics to handle the extremely short time scale between collisions. These are two most challenging issues facing the inner detector, and the choice of materials and machinery is chosen specifically to handle these issues.

The inner detector covers roughly $|\eta| < 2.5$, and is further subdivided into three independent, yet complementary subdetectors: the pixels, the semiconductor tracker (SCT) and the transition radiation tracker (TRT). The whole inner detector is cylindrical in shape, with length of 6.2 m, and diameter of 2.1 m and is pictured in Fig. 4.3.

The Pixel Detectors

The pixel system is the innermost subdetector of ATLAS. In order to have great ability in recognizing between primary and secondary vertices, the pixel detector has very high granularity, which allows for high precision track measurements. The pixel system has intrinsic transverse precision of 10 μm and longitudinal precision of 115 μm.

The system consists of three cylindrical layers of overlapping silicon pixel sensors at transverse positions of roughly 5, 9 and 12 cm from the beam line, as well as three perpendicular disk layers positioned between 50 and 65 cm from the interaction point. There are a total of 1744 identical pixel sensors, each with 460,080 readout channels for the individual pixels. The pixels themselves each have a nominal size of 50×400 μm^2.

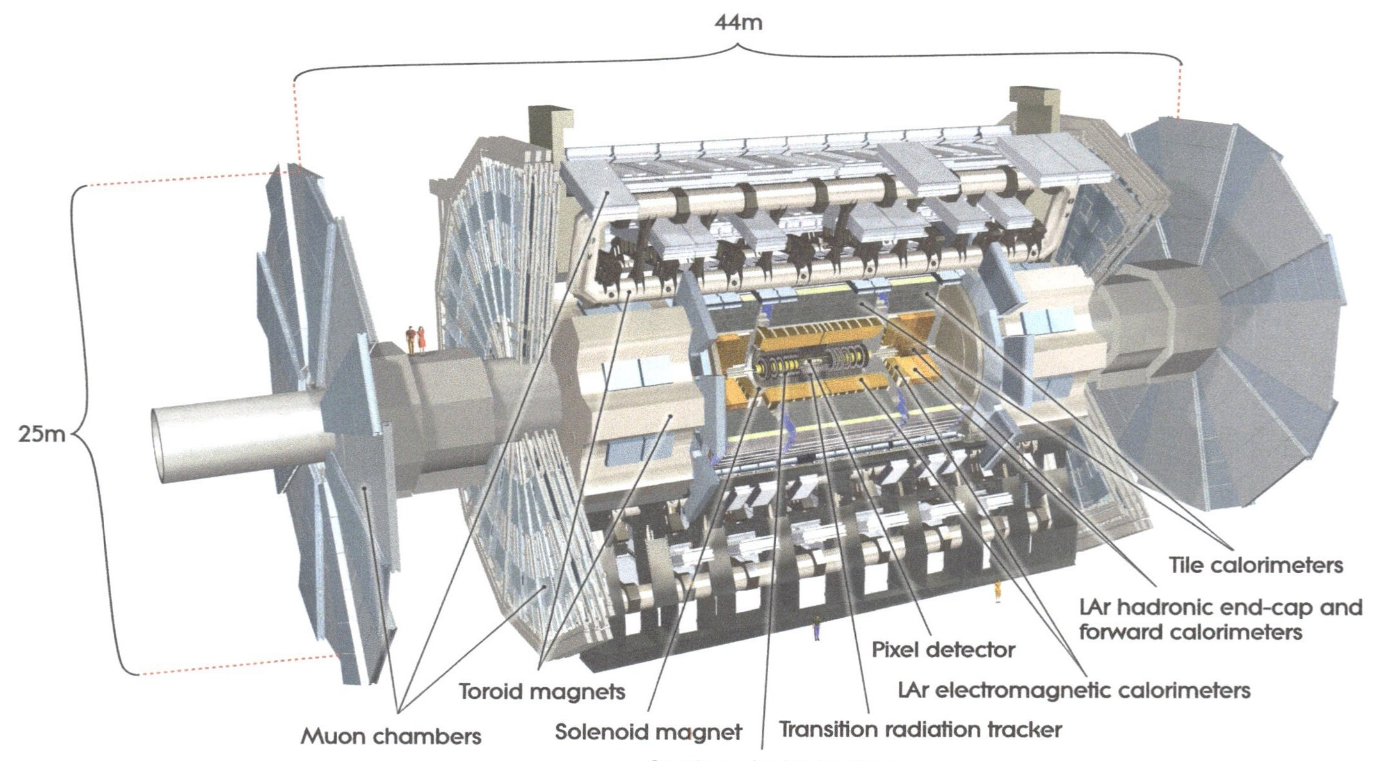

Fig. 4.2 Cutaway view of the entire ATLAS detector. Taken from [41]

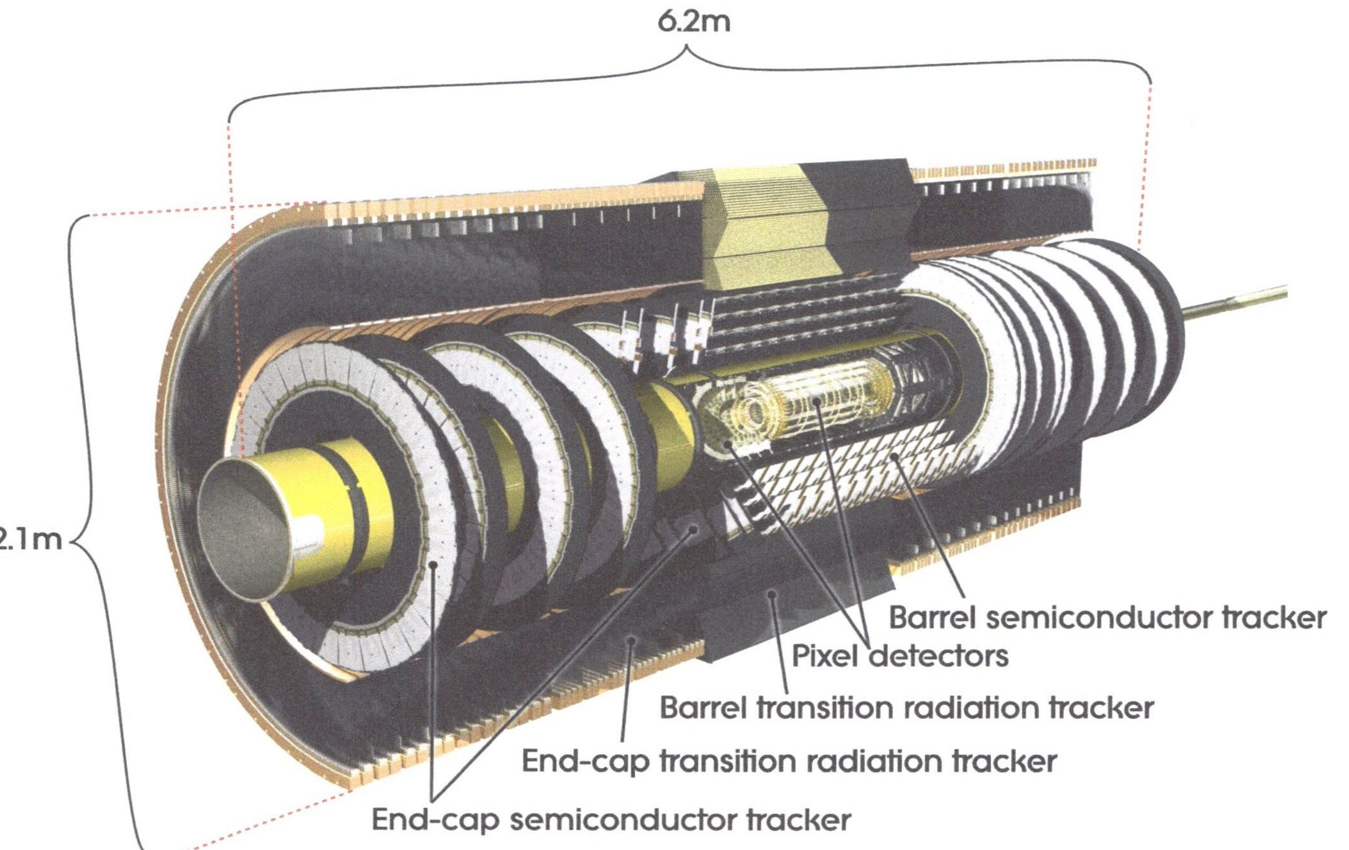

Fig. 4.3 Cutaway view of the inner detector with all subdetectors labelled accordingly. Taken from [41]

The Semiconductor Tracker

The SCT [42] is similar to the pixel system in that it is made up of cylindrical layers of sensors going radially out from the beam line, as well as disk shaped sensors positioned on either side of the interaction point. There are a total of four cylindrical layers, located between 30 and 50 cm from the beam line, and nine layers of disks on either side of the interaction point, located between 85 and 270 cm from the centre.

The sensors themselves are made of microstrip silicon and are arranged in pairs, with one sensor parallel to the beam line (in the case of the cylindrical barrel layers) and the other at a 40 mrad angle. This, along with a 80 μm strip pitch, allows for an intrinsic transverse precision of 17 μm, and a longitudinal precision of 580 μm.

The Transition Radiation Tracker

The TRT [43, 44] is the outermost layer of the inner detector. It uses a very different detection technology to measure tracks from charged particles. The TRT is made up of millions of drift (straw) tubes of 144 cm for the barrel region, and 37 cm for the end-caps. They are made of polyimide film, which acts as the cathode, and gold plated tungsten wires centered in the middle, which act as anodes. The straws are filled with a $Xe/CO_2/O_2$ gas mixture. In the barrel region, the straw tube modules are located between 55 and 110 cm from the beam line, and cover a region of $|\eta| < 1.0$. The end-caps are located between 80 and 270 cm up or down from the interaction point, and additionally cover $1.0 < |\eta| < 2.0$.

The TRT has an intrinsic transverse precision of 130 μm, which is much larger than the other two detectors. However, it does have the ability to identify different particles. As charged particles pass through the straws, transition radiation photons are emitted and detected by the straw tubes. The number of radiated photons is roughly proportional to the particle mass, thus allowing the distinction between electrons and heavier particles.

4.2.2 The Calorimeters

When a high energy particle interacts with matter, a cascade of lower energy particles (also known as a particle shower) is produced. The energy of the original particle is split among the secondary particles produced, which can further be split among tertiary particles, and so on, until there are several low energy particles, that are absorbed by the material. The calorimeter system, which is located around the inner detector and the solenoid, has as primary task to measure the energy of particles that produce these cascades.

The ATLAS calorimeter system covers the region $|\eta| < 4.9$, which is very important to contain as many particle showers as possible and to accurately measure the missing transverse energy. Similar to the inner detector, the calorimeters are divided into both a cylindrical barrel region and disk shaped endcaps. The shape and depth of particle showers differ depending on the original particle type; therefore, the calorimeter system is also divided into two major layers based on the two particle shower types: EM and hadronic [5]. The system is pictured in Fig. 4.4.

The Liquid Argon Electromagnetic Calorimeter

EM showers occur when a high energy electron or photon interacts with a material. A high energy photon interacts with matter primarily through pair production, where an e^+e^- pair is produced, splitting the photon's energy between the two particles. Electrons (and positrons) interact primarily through a process called bremsstrahlung, in which the electron (or positron) is slowed when it emits a photon. These two processes alternate for any high energy EM particle as it passes through matter, until the resulting particles have low enough energy to be completely absorbed by the material through another process, such as ionization (for electrons and positrons) and the photoelectric effect (for photons). The amount of matter traversed by a particle is measured in units of radiation length, X_0 (which have units of g$\cdot$cm^{-2}). One radiation length is characterized by (a) the mean distance over which a high energy electron loses all but $1/e$ of its energy, and (b) 7/9 of the mean free path for pair production by a high energy photon. Radiation length is a material dependent value, with higher electron density leading to a lower radiation length, thus reflecting a better stopping power. The depth of the EM shower is directly related to the initial particles energy, which allows for it to be measured.

The Liquid Argon (LAr) EM calorimeter is divided into two components: a barrel region ($|\eta| < 1.475$) and two endcaps ($1.375 < |\eta| < 3.2$). The barrel is divided into two identical halves that are separated by a small gap (4 mm) at $z = 0$. Each half barrel is 3.2 m long, and has an inner (outer) radius of 1.4 m (2.0 m). Each endcap is divided into two coaxial wheels: an outer wheel covering the region $2.5 < |\eta| < 3.2$, and an inner wheel, covering $1.375 < |\eta| < 2.5$. The barrel is segmented into three sections in depth, while the endcaps are segmented into two sections.

Particle showers are measured by alternating dense inactive layers, with less dense active layers that detect energy deposition[1]. Liquid argon is used as the active material for the LAr calorimeters, with accordion shaped kapton electrodes and lead absorber plates, which allows for full coverage in ϕ. The thickness of the lead plates was optimized as a function of η in order to enhance the energy resolution.

The granularity of each individual segment is dependent on its location. In both the barrel and the endcaps, the granularity ranges from very fine segmentation in η, $\Delta\eta \times \Delta\phi = 0.025/8 \times 0.1$, to fine overall segmentation, $\Delta\eta \times \Delta\phi = 0.025 \times 0.025$. This design is used to ensure that desired resolution is achieved without costing too much. In general, the middle layer has the finest granularity, since the majority of energy deposition is expected to be there.

[1] This alternating of active and inactive layers is called a sampling calorimeter.

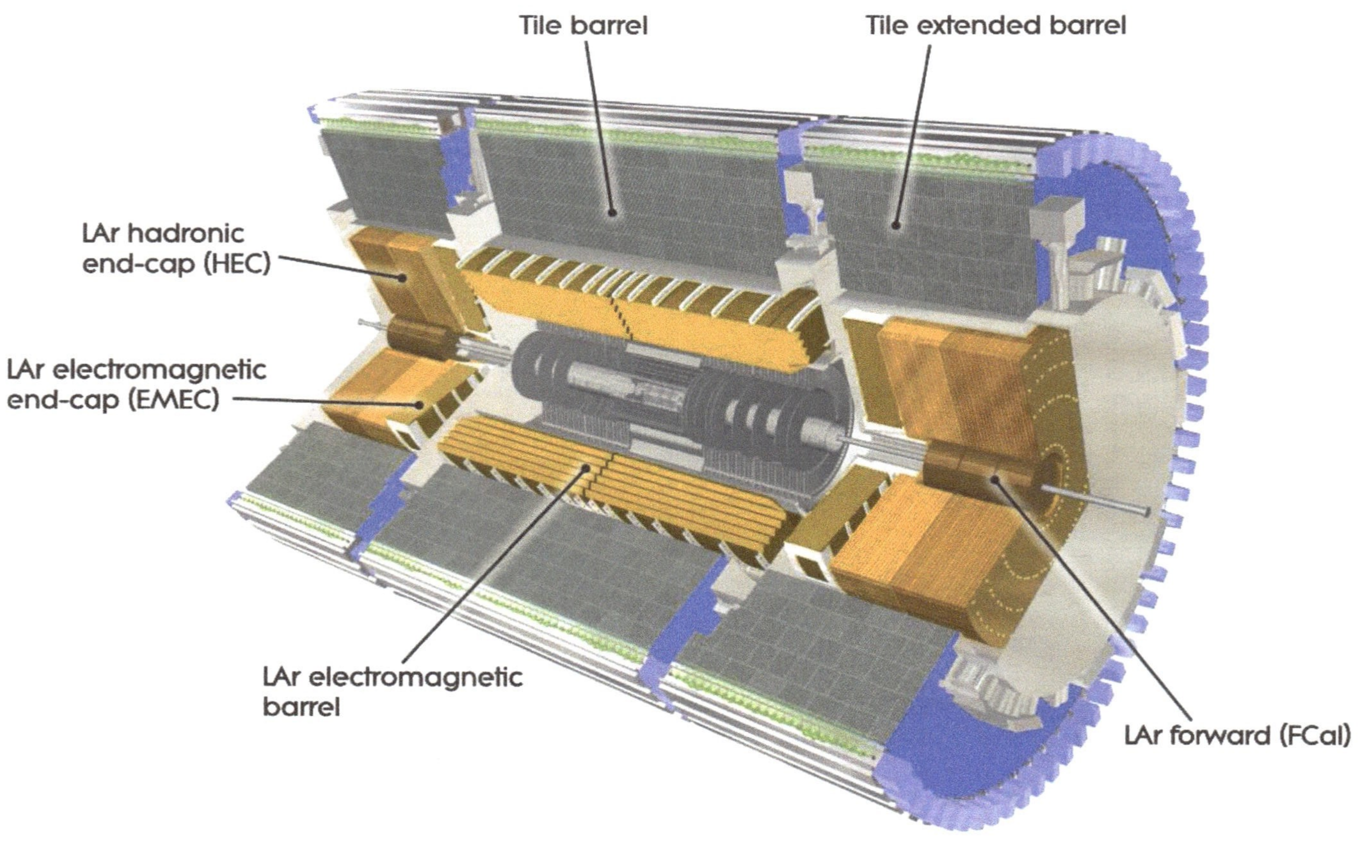

Fig. 4.4 Cutaway view of the calorimeter system, with all subdetectors labelled accordingly. Taken from [41]

Hadronic Calorimeters

Hadronic showers occur when high energy hadrons interact with matter through the strong force. This occurs when a high energy hadron interacts with the nuclei of the given material, and splits into hadrons of lower energy. This continues until the resulting particles have low enough energy to be stopped by the material. However, there is an added complexity since hadronic interactions can often produce EM particles (through the decay of a neutral pion to two photons, for example), which can initiate an EM shower if they are of high enough energy. The fraction of EM particles produced in a hadronic cascade can change significantly from interaction to interaction, making energy resolution difficult. In addition, lost energy through processes such as nuclear excitation and breakup do not leave detectable signals, which can further lead to lower energy resolution. Hadronic showers do take longer to develop, requiring more material to fully contain. Like EM showers, hadronic showers are measured in terms of nuclear interaction lengths, λ_I, which is the hadronic analogue to X_0.

The hadronic calorimeters are placed directly behind the EM calorimeters and are divided into both a barrel and endcaps. The barrel is further divided into three sections: a central barrel ($|\eta| < 1.0$) and two extended barrels ($0.8 < |\eta| < 1.7$). The barrels all have an inner (outer) radius of 2.28 m (4.25 m), and has a total length of 11 m. The wheel-shaped endcaps overlap slightly with the barrel calorimeters and cover the range $1.5 < |\eta| < 3.2$. Each endcap has two independent wheels, that are built from 32 identical wedge-shaped modules, each with two segments. This amounts to four layers per hadronic endcap.

The hadronic endcaps use liquid argon as an active material, with copper absorber plates, and share the same cryostat as the LAr EM endcap calorimeters. The barrel, however, uses scintillating tiles as its active material, and steel absorber plates.

The granularity of the hadronic calorimeters is coarser than the EM calorimeters. The tile calorimeter has module sizes of $\Delta\eta \times \Delta\phi = 0.1 \times 0.1$ except for the last layer, which has sizes of $\Delta\eta \times \Delta\phi = 0.2 \times 0.1$. The hadronic endcaps have similar granularity, but dependent on η, with the region $1.5 < |\eta| < 2.5$ having granularity of $\Delta\eta \times \Delta\phi = 0.1 \times 0.1$, while $2.5 < |\eta| < 3.2$ has $\Delta\eta \times \Delta\phi = 0.2 \times 0.2$.

LAr Forward Calorimeter

In the very forward region, there is also the LAr Forward Calorimeter (FCal). This covers the region of $3.1 < |\eta| < 4.9$. It shares the same cryostat as the hadronic and EM calorimeters, but is recessed slightly (1.2 m) with respect to the inner EM endcap. Because of this, the FCal must be much denser than the other calorimeters. Liquid argon is still used an active material, but both copper (for the first layer) and tungsten (for the next two layers) are used as absorbers. The granularity is still quite small, with module sizes ranging from $\Delta x \times \Delta y = 3.0/4$ cm $\times$ $2.6/4$ cm in the first layer to $\Delta x \times \Delta y = 5.4$ cm $\times$ 4.7 cm in the last layer.

4.2.3 The Muon System

Most particles created at the interaction point are contained within the calorimeters. There are two exceptions to this: neutrinos, which cannot be detected at ATLAS, and muons, which are often low-ionizing and able to pass through the entire calorimeter without losing too much energy. Measuring muon energies and momenta are major design goals of ATLAS, and thus surrounding the calorimeters is a multi-staged muon system, pictured in Fig. 4.5.

Three different types of muon detectors are used to measure muon tracks: Monitored Drift Tubes (MDT's), which cover the region $|\eta| < 2.7$, Resistive Plate Chambers (RPC's), covering $|\eta| < 1.05$, Thin Gap Chambers (TGC's), covering $1.05 < |\eta| < 2.7$, and finally Cathode Strip Chambers (CSC's), covering $2.0 < |\eta| < 2.7$.

Muon flight paths are bent by one of three air core toroid magnets. These are made of eight rectangular shaped coils arranged in a circle along the beam line. The central barrel toroid is the largest of the three, covering the region $|\eta| < 1.4$ and producing a magnetic field of 0.5 T. Two identical endcap toroids covers the region $1.6 < |\eta| < 2.7$ and produce magnetic fields of 1 T. The region in between the two is called the transition region, where muons still feel a magnetic field, but with contributions from both toroids. The reconstructed muon tracks are used to measure the muon's momentum.

4.3 Data Preparation

As mentioned in Sect. 4.1, the minimum spacing between bunches at the LHC is 25 ns. This corresponds to a maximum bunch crossing rate of 40 MHz. With the amount of information collected from all the various subdetectors, it would be nearly impossible to collect data at this rate. A tiered triggering system is in place to reduce the data collection to a manageable rate. The quality of the collected data must meet the highest standards in order for it to be fit for physics analyses. These topics and more are covered in this section.

4.3.1 Triggering

The ATLAS trigger is a three-tiered system, tasked with reducing the data collection to a manageable rate (about 200 Hz). The three levels are Level 1 (L1), Level 2 (L2) and Event Filter (EF). L1 is a hardware-based trigger, whereas L2 and EF, often grouped together as the High Level Trigger (HLT), are a software-based trigger.

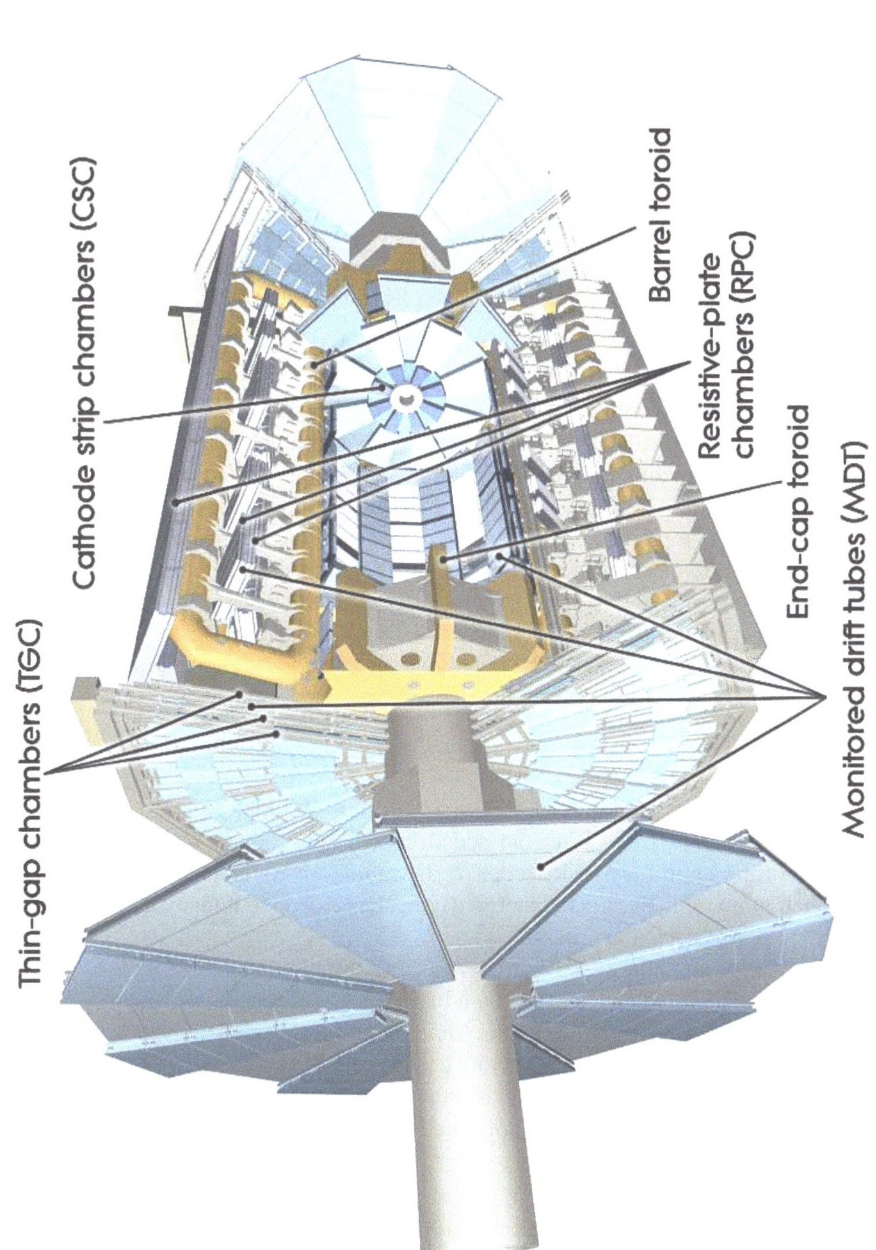

Fig. 4.5 Cutaway view of the muon system, with all subdetectors labelled accordingly. Taken from [41]

Level 1 Trigger

The L1 trigger reduces the ATLAS event rate from 40 MHz[2] to 75 kHz. This is accomplished by searching for events with high transverse momentum muons, electrons, photons, jets and τ-leptons decaying to hadrons, as well as events with large missing transverse energy, or large total transverse energy. Different energy thresholds and multiplicities can be used in various combinations, but the number of individual triggers is limited by memory.

Trigger decisions at L1 must take less than 2.5 μs, thus memory and CPU usage is limited and continuously monitored to stay within this time constraint. This is accomplished partly by only using a subset of detectors, each with reduced granularity.

The identification of the different particles is very rudimentary, mainly looking for a cluster (or a group of clusters) of energy deposition meeting some threshold. However, each particle location is saved and used as a seed at L2.

Level 2 Trigger

The L2 trigger reduces the ATLAS event rate from 75 kHz to less than 3.5 kHz. As mentioned above, particle information from L1 is stored and used as a seed at L2. This reduces the amount of detector data transferred to the L2 computers.

Trigger decisions at L2 must take less than 40 ms. This added time, compared to L1, allows for more detector information to be used for event selection by having greater granularity per subdetector. This allows for more sophisticated algorithms to be used when identifying various particles, such as distinguishing between electrons and photons. Similar to L1, events must satisfy at least one of a predetermined list of trigger signatures in order to be kept. If an event satisfies this requirement, it is passed on to the next trigger level.

Event Filter Trigger

The EF trigger reduces the ATLAS event rate from 3.5 kHz to 200 Hz, the data recording rate at ATLAS. With the added latency time (about four seconds) full detector information and a wider scope of algorithms are available at EF. This allows for much more refined trigger signatures to be developed, with many EF thresholds matching offline selection criteria.

Events kept after the EF are subdivided into different streams based on which trigger fired. The streams are basically a list of EF trigger chains and any event that fires at least one is put into the stream. Streams are inclusive, meaning an event can be in multiple streams. These streams are used for one of two purposes: calibration or physics, allowing for much less redundancy when running on data and only needing it for either purpose. Not all event information is kept for calibration streams, since

---

[2] Note, this is a maximum possible event rate, based off the 25 ns bunch crossing. In reality, bunch crossing rates are typically longer, and not all bunches interact.

only certain data is relevant. All event information is stored for physics streams, and are further subdivided into general physics chains, eg. mainly EM signatures (EGamma stream), or events with jets, missing energy or taus (JetTauEtmiss stream). In addition, an express stream with only about 10 % of the data is written out to help with data quality.

4.3.2 Computing Model

The complexity of the computing model [45] matches that of the triggering system. The size of events that pass the EF trigger is roughly 1.6 MB in its RAW format, with all detector information stored. These events are then sent through reconstruction and calibration algorithms to identify the particles in each event and store the information in an object-oriented representation. This reduces the event size by removing most of the detector specific output and leaving mostly information on reconstructed objects. This is done in two steps, first to an Event Summary Data (ESD) format, then to an Analysis Object Data (AOD) format. Most analyses can be done directly from the AOD using Athena (the ATLAS software framework) and/or ROOT [46].

The steps mentioned above are very complex and require considerable amounts of CPU usage, and memory storage capacity. To help alleviate the load on the facilities at CERN, a tiered computing model is used to allocate tasks to other computing centres. The Tier-0 facility is located directly at CERN and is responsible for storing and distributing the RAW data directly from the EF trigger. All prompt reconstruction of event streams occur at Tier-0, which allows for easy availability and fast response times for calibration and data quality groups. There are 10 Tier-1 facilities located worldwide, with dedicated links to CERN. Each Tier-1 facility stores a subset of the RAW data, and is responsible for the reprocessing of the data, which is typically a longer process than the prompt reconstruction at Tier-0. Tier-1's are also responsible to provide collaboration-wide access to the analysis ready data formats (ESD, AOD). The role of the ~ 50 Tier-2 facilities is fairly broader than the previous two. These facilities host subsets of derived data formats, perform simulations and analyses for working groups, and perform specific calibrations, based on local interest. Finally, Tier-3 facilities are located at most institutions, and are primarily dedicated to analyses of local interest, and the storage of small subsets of data for the purpose of end-user work.

4.3.3 Data Quality

During normal operation of the ATLAS detector, data from some component may be unusable for physics studies during a certain period of time. This can be due to components being at non-nominal voltages, readout electronics needing to be reset, or data being noisier than usual. Data acquisition continues despite these. To ensure the integrity of the data collected at ATLAS, systems are in place to first identify data that may be at risk of being of lower quality, then, depending on the severity, either fix or exclude the event [47].

Data collected at ATLAS are divided into runs. These are defined as all data collected during one particular fill of the LHC, which nominally lasts 12 h, but can often be shorter or longer. To help identify problems at ATLAS, each run is further subdivided into small manageable units of time called Luminosity Blocks (LBs). The exact length is dependent on the luminosity of the LHC, but for 2011 data, they were roughly one minute, two minutes in 2012. During data collection, a small subset of data (the express stream, mentioned above) is initially reconstructed and monitored. Through automatic checks on filled histograms, and expert input, problems in the detector are located and identified. If the problem is simple, eg. a noisy cell, the noisy cell is removed, and not included in the full reconstruction. However, if the problem is larger, and affects an entire LB (or more), each affected LB is assigned a defect describing the problem. Thirty-six hours after each run, and after this initial monitoring has occurred, reconstruction begins on the entire dataset, with any problems resolved. A second round of assessments occur once this is complete to ensure that all problems are resolved.

The result of this is a Good Run List (GRL), a list of runs and LBs that are suitable for physics analysis. Several different GRLs exist, each geared towards a particular type of analysis, requiring certain components of the detector to be in good working condition, but not necessarily others. This is very important because little data is lost due to a problem in a component that does not affect a given analysis. GRLs are also used to determine the integrated luminosity over a given period of time. This is accomplished by integrating the instantaneous luminosity over the individual LBs that are included in a GRL. This is essential, since instantaneous luminosity changes over time, as mentioned in Sect. 4.1.1.

4.3.4 Data Collected in 2011

The data used in this analysis was collected in 2011. The peak luminosity reached during this time was $3.65 \times 10^{33}\ \mathrm{cm}^{-2}\mathrm{s}^{-1}$. The evolution of the peak instantaneous luminosity per run is plotted in Fig. 4.6.

This amounted to $5.25\ \mathrm{fb}^{-1}$ ($5.61\ \mathrm{fb}^{-1}$) of recorded (delivered) integrated luminosity over the entire year, as plotted in Fig. 4.7. The most data recorded in a single day was $128.9\ \mathrm{pb}^{-1}$.

4.4 Definition of Physics Objects

The identification and measurement of particles passing through the ATLAS detector are essential for conducting physics analyses using data collected. Fortunately, each type of particle reacts slightly differently to the various subdetectors, leaving a

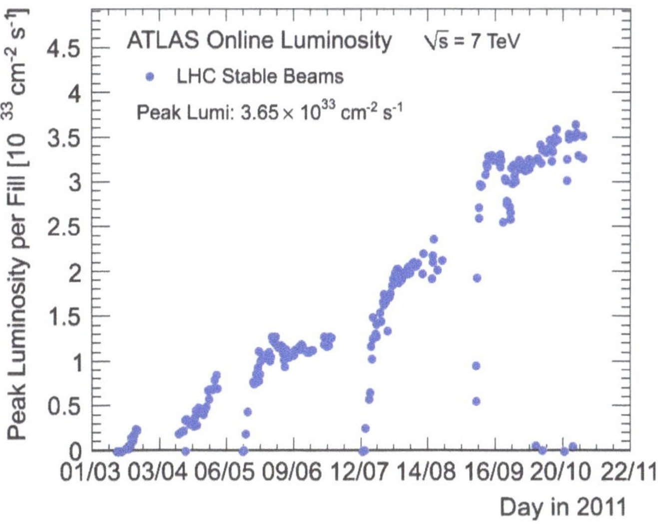

Fig. 4.6 Peak instantaneous luminosity per run of the LHC during 2011 data taking period. Taken from [48]

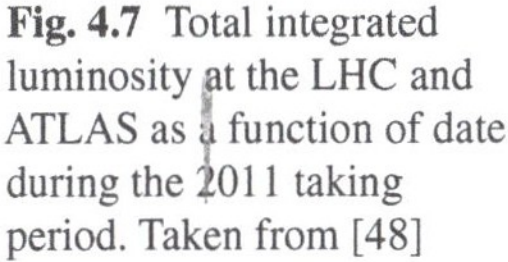
Fig. 4.7 Total integrated luminosity at the LHC and ATLAS as a function of date during the 2011 taking period. Taken from [48]

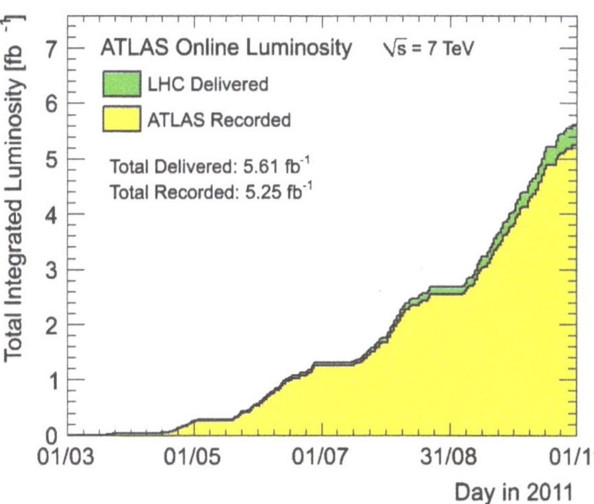

unique signature identifying it, as shown in Fig. 4.8. For example, charged particles, such electrons, protons and muons, leave curved tracks in the inner detector, while depositing the majority of their energy in the EM calorimeter, hadronic calorimeter and muon detectors, respectively. Uncharged particles, such as photons and neutrons, do not have bent paths in the inner detector, but instead leave the majority of their energy in the EM and hadronic calorimeters, respectively. The weakly interacting neutrinos travel through the machine undetected, leaving a signature of missing energy in the detector.

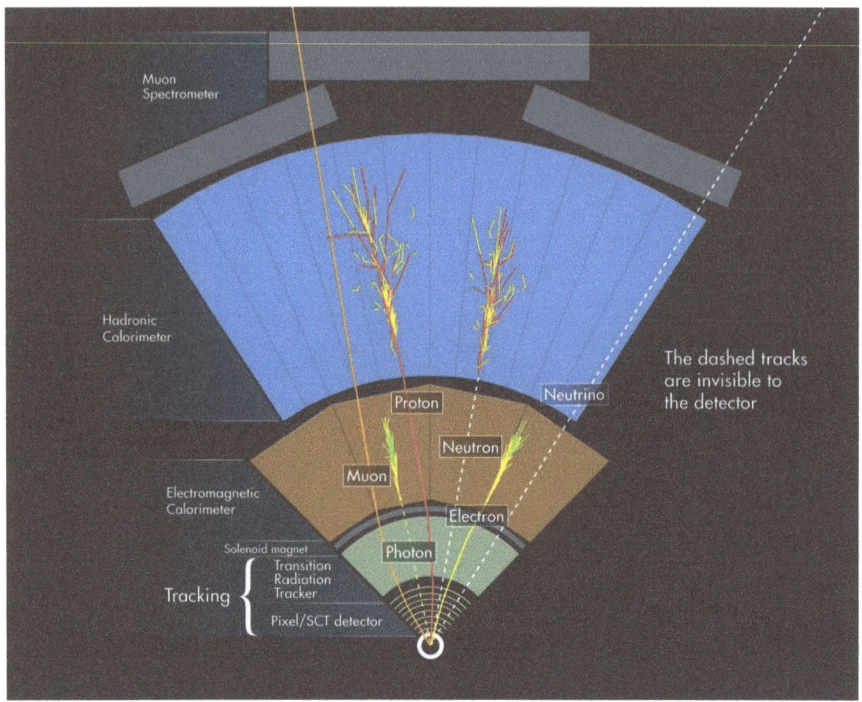

Fig. 4.8 Cross sectional view of the ATLAS detector showing the passage of various particles passing through the detector. Taken from [41]

Each particle has its own reconstruction algorithm(s) to identify those particles within the detector. Many studies have been conducted to determine the efficiency and performance of these algorithms to not only identify, but correctly measure the energy and momentum of these particles. Only jets are considered in this analysis, thus only those relevant algorithms are presented here.

4.4.1 Jet Identification and Reconstruction

As mentioned in Sect. 2.1.2, due to confinement, partons (quarks and gluons) cannot exist by themselves in nature. Thus if a parton is produced or scattered during a collision at the LHC, as it travels through the detector, it will immediately hadronize: the spontaneous creation of hadrons to compensate for the free parton. This appears as a cascade or shower of hadronic particles within a narrow cone, all in the direction of the original parton. This is commonly known as a jet. Though the original parton is not directly detected, its energy can be determined by combining the four-momenta of these particles; the resulting energy and momentum should be similar to that of the original parton.

There are two types of inputs for jet search algorithms: topological clusters or towers of cells. Clusters are three dimensional regions of energy deposition, seeded by cells with large signal significance (compared to noise), with neighbouring cells' energy added. Towers are made up of cells that fit within a fixed size ($\Delta\eta \times \Delta\phi = 0.1 \times 0.1$), and have energy equal to the sum of those cells. These inputs are objects used in two main types of algorithms that attempt to reconstruct the energy and momentum of jets: the cone and cluster algorithms [49].

In the cone algorithm, an object above some energy threshold is used as a seed in a process that combines the objects that are a fixed distance from the seed (ie. within a cone around the seed). This is an iterative process, with the momentum of the combined objects recalculated after each iteration and used as the centre of the next cone. This is done until a stable momentum is achieved.

The cluster algorithm uses a completely different principle. Objects are combined until a certain condition on the distance and energy of these objects are met. For most cluster algorithms, a list is compiled of the following terms:

$$d_{ij} = \min\left(k_{Ti}^{2p}, k_{Tj}^{2p}\right) \frac{(\Delta R)_{ij}^2}{C^2} \tag{4.3}$$

$$d_{ii} = k_{Ti}^{2p} \tag{4.4}$$

where i, j are over all objects, k_{Ti} is the transverse momentum of the ith term, ΔR_{ij} (defined in Eq. 2.10) is the angle difference between the two objects, p is some number, depending on the specific algorithm used, and C is an angle difference parameter of the algorithm, which sets the resolution of the jets. Once the list is compiled, the minimum value is found. If d_{ii} is the minimum, then that object is removed from the list and put into the jet container. If the minimum is d_{ij}, then the i and j objects are combined together, using some clustering scheme and the new object is put back into the list. The values of (4.3) and (4.4) are recalculated and the process begins again. The most common values of p are 1 (for the k_T algorithm), 0 (for the Cambridge algorithm) and -1 (for the Anti-k_T algorithm).

The accurate reconstruction of jets is challenging, since one is always building a jet from smaller composite objects. In choosing a jet algorithm, two major guidelines must be met for it to be trusted: infrared and collinear safety. For infrared safety, the presence (or absence) of low energy, soft particles between objects of a jet must not have an effect on the recombination of these particles into a jet. For collinear safety, there should be no difference in reconstruction if one has a single high momentum particle, or a particle of the same momentum split into two collinear particles. The AntiKt algorithm [50] has proven to be most successful at satisfying these two conditions. Thus it is the standard jet algorithm used in ATLAS. In addition, topological clusters are the standard in ATLAS jet algorithms, since they fare much better with pile-up compared to the towers.

The energy of jets [51] is measured by combining the energies of its individual components, the clusters. These clusters are reconstructed at the EM scale, meaning

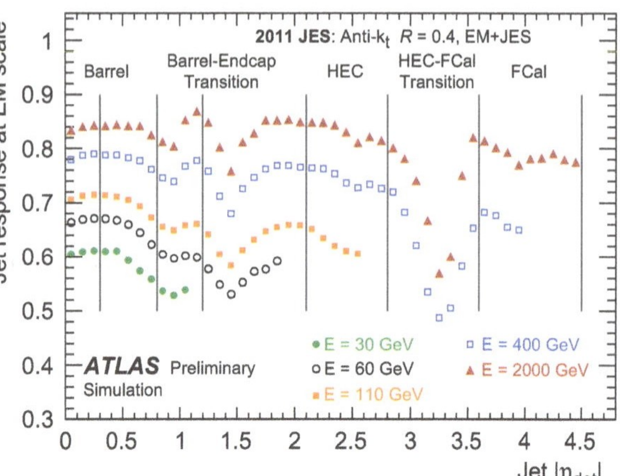

Fig. 4.9 Average jet response ($E_{jet}^{data}/E_{jet}^{truth}$) for jets derived from EM calibrated topological clusters as a function of η, for various jet energies. Note that results are based on Pythia MC samples. Taken from [51]

they can correctly measure the energy deposited by EM showers in the calorimetry (due to beam tests being conducted on the calorimeters prior to the ATLAS construction using electron beams). Several corrections are made to jets after reconstruction such that they have the correct Jet Energy Scale (JES). These include a pile-up correction, to account for the energy offset caused by additional pile-up interactions, a jet direction correction, in order to point the jet in the direction of the primary vertex[3] of the event, a Monte Carlo (MC) derived correction to match reconstructed jet energies to their truth level[4] energies (see Fig. 4.9 to see this jet response value as a function of η for different values of jet p_T), and finally a residual *in situ* derived correction, which exploits the p_T balance between a jet and a reference object (another jet, a photon, or a Z), matching both data and MC samples.

The JES carries its own uncertainty, which is derived from the absolute and the relative *in situ* calibration uncertainties being added in quadrature. The uncertainty in the JES is dependent on both η and p_T^{jet}, but is less than 1 % for jets with $55 \leq p_T^{jet} < 500$ GeV, and $|\eta| < 1.2$. This increases slightly for forward jets ($|\eta| > 1.2$), reaching as high as 6 % for very forward, and low p_T jets. Generally though, the uncertainty decreases with increasing p_T. Uncertainties in JES are plotted in Fig. 4.10

The development of jet reconstruction algorithms was accomplished by working groups within the collaboration, and the software used to determine the JES was maintained centrally by the same working group.

[3] Several interactions can occur simultaneously during a bunch crossing at ATLAS, leading to "pile-up" events. To distinguish between them, primary vertices are determined, which are vertices with the largest total p_T of tracks associated to them.

[4] The energy of a jet at "truth level" corresponds to the energy of the parton from which the jet originated. Truth level energies are not known during data taking; however, they are known when conducting MC studies, as the energy of all particles are quantified during MC generation.

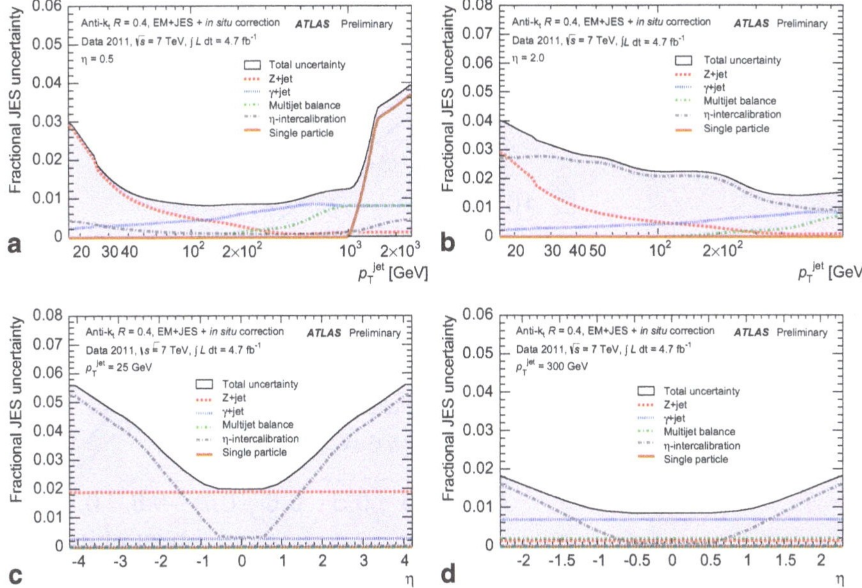

Fig. 4.10 Plots of fractional JES uncertainty as a function of p_T^{jet} at (**a**) $\eta = 0.5$ and (**b**) $\eta = 2.0$, and plots of fractional JES uncertainty as a function of η for (**c**) $p_T^{jet} = 25$ GeV and (**d**) $p_T^{jet} = 300$ GeV. Total uncertainty is plotted in black, which is a combination of several different components (also plotted), added in quadrature. Taken from [51]

4.4.2 *b*-tagging Algorithms

A very important aspect to jet reconstruction is the distinction of jets arising from b-quarks to those arising from lighter quarks (u's, d's, s's or c's) or gluons. The identification of b's is very important for ATLAS since they are found in many final state decays, for example the decay of the SM Higgs to $b\bar{b}$.

The decay and hadronization of b-quarks are different compared to the lighter quarks and gluons, which can allow for b-quarks to be identified. First, the b-hadron produced from the b's hadronization tends to have a large percentage of the initial momentum (about 70 %), resulting in a harder fragmentation. Second, the resulting hadron has relatively high mass ($>$ 5 GeV), thus its decay products tend to have large transverse momentum with respect to the jet axis. Finally, the most important property, the lifetime of b-hadrons is relatively long ($\sim$ 1.5 ps), which for a 50 GeV b-quark leads to a flight path of roughly 3 mm. Fortunately the inner detector has very good spatial resolution, which allows for the distinction of displaced vertices.

There are several algorithms that can be used to help identify jets arising from b-quarks (or b-jets for short). They are all spatial in nature, and utilize one or more of the properties of b-jets described above [52]. The IP3D algorithm utilizes impact

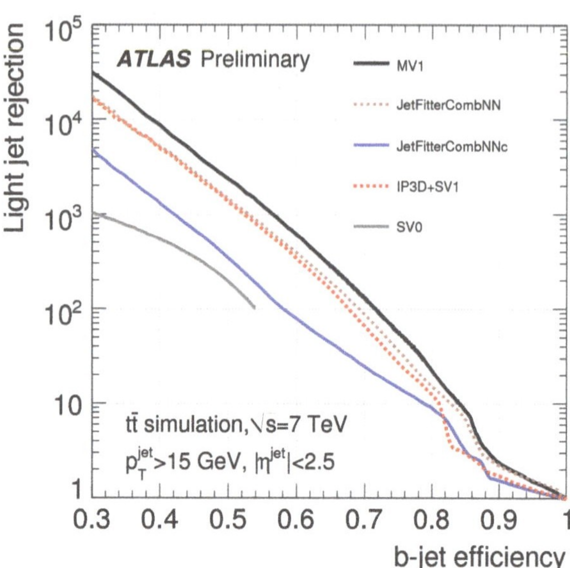

Fig. 4.11 Plots of the b-tagging efficiency versus light-jet rejection for various b-tagging algorithms used at ATLAS. Taken from [53]

parameter significance of tracks within the jet, SV0 and SV1 use displaced vertex information, JetFitter also uses displaced vertex information and b-quark flight length. To better discriminate between b- and light-jets, these various algorithms can be combined together in some way, often using a neural network. The primary algorithm used for this analysis is JetFitterCOMBNN, which combines the results from JetFitter and IP3D using a neural network.

The success of any b-tagging algorithm is often determined by two numbers: the b-tag efficiency (the rate at which a b-jet is actually b-tagged) and the light-jet rejection[5] (the number of light-jets that are properly tagged for every light-jet that is misidentified as a b-tagged jet). These two quantities are typically inversely proportional to each other, with one increasing as the other decreases. This is logical, since one can choose a b-tag working point that is very efficient at picking out b-jets, but it will inevitably misidentify more light-jets as b-tagged. Plots of b-tag efficiency and light-jet rejection for various algorithms are shown in Fig. 4.11 based on $t\bar{t}$ simulated events. The working point used for this analysis is the 57 % JetFitterCOMBNN[6] b-tag efficiency, which amounts to a light-jet rejection of ~ 600.

[5] Light-jet rejection is often referred to as mistag rate, or light-jet efficiency, which is simply the inverse of light-jet rejection.

[6] A b-tagging algorithm developed and maintained centrally by a flavour tagging working group at ATLAS.

Table 4.1 Cross sections and branching ratios for the VBF production of Higgs bosons and the $H^0 \to b\bar{b}$ decay channel for $\sqrt{s} = 7$ TeV

m_H [GeV]	σ_{VBF} [pb]	$BR_{H^0 \to b\bar{b}}$	$\sigma \times BR$ [pb]
115	$1.344^{+2.5\,\%}_{-2.3\,\%}$	0.703	0.9377
120	$1.279^{+2.7\,\%}_{-2.5\,\%}$	0.648	0.8223
125	$1.222^{+2.8\,\%}_{-2.4\,\%}$	0.577	0.6987
130	$1.168^{+2.8\,\%}_{-2.3\,\%}$	0.494	0.5689

4.5 Monte Carlo Methods and Samples

A Monte Carlo (MC) method is a technique that uses pseudorandom numbers to help solve a problem using a probabilistic approach. This is often done by using known (or expected) probability distributions for a given input, randomly generating inputs using these distributions, and investigating the final outcome in numerous pseudo-experiments. This can then be compared to the actual final outcome in the given experiment.

At ATLAS, MC methods are most often used in two cases: generating events arising from proton-proton collisions for a given process (as per the expected results from its Feynman diagram), and simulating the passage of the final state particles of the events through the actual physics detector.

Event generation is accomplished by using several input variables, such as PDFs, cross sections and hadronization models. There exist several event generators, each having their own strengths and weaknesses. For this analysis, Pythia [54], Alpgen [55], Herwig [56], Jimmy [57], and MC@NLO [58] are all used. Pythia, Herwig and Alpgen all generate events using lowest order calculations, MC@NLO uses NLO calculations, and Jimmy is used for the hadronization of multi-parton interactions.

Once the events have been generated, the passage of the particles through the detector is modelled using GEANT4 [59]. The electronic signal expected from this energy deposition is then used to create output files, that are in the same format as those created from real data, such that they can be analyzed in a similar fashion.

4.5.1 Monte Carlo Samples Used in this Analysis

Signal

The VBF $H^0 \to b\bar{b}$ signal MC samples used for this analysis were generated using Herwig, with no generator level filter applied. Four Higgs mass points are considered in this analysis: $m_H = 115, 120, 125, 130$ GeV. The production cross section and the branching ratio are calculated centrally by the LHC Physics Higgs Cross Section group [20], with values listed in Table 4.1.

Table 4.2 Cross sections for Pythia dijet samples used in this analysis. Each sample is divided into 'dijet p_T' slices (Jn)

Sample	p_T range [GeV]	σ [pb]
Pythia Dijet J0	8–17	1.20×10^{10}
Pythia Dijet J1	17–35	8.07×10^{8}
Pythia Dijet J2	35–70	4.80×10^{7}
Pythia Dijet J3	70–140	2.54×10^{6}
Pythia Dijet J4	140–280	9.96×10^{4}
Pythia Dijet J5	280–560	2.59×10^{3}
Pythia Dijet J6	560–1120	35.5
Pythia Dijet J7	1120–2240	0.134
Pythia Dijet J8	2240–∞	5.68×10^{-6}

Table 4.3 Cross sections for Alpgen multijet samples used in this analysis. Slices refer to leading jet p_T (identical to p_T ranges described in Table 4.2). Lower slices have a generator-level filter applied to them, requiring 4 jets with $p_T > 17$ GeV and 3 jets with $p_T > 25$ GeV. All numbers are in pb, and each slice is divided into number of primary partons in the final state (2 to 6)

Slice	2 partons	3 partons	4 partons	5 partons	6 partons
J1 (4j17_3j25)	1.11×10^{5}	2.34×10^{4}	–	–	–
J2 (4j17_3j25)	7.54×10^{4}	1.70×10^{5}	2.67×10^{5}	5.87×10^{4}	9.82×10^{3}
J3 (4j17_3j25)	1.15×10^{4}	1.17×10^{5}	1.74×10^{5}	5.42×10^{4}	1.38×10^{4}
J4	1.57×10^{4}	2.95×10^{4}	2.01×10^{4}	8.70×10^{3}	3.42×10^{3}
J5+	325	754	710	421	241

Table 4.4 Cross sections for Alpgen $b\bar{b} + n$ parton samples used in this analysis. Slices refer to leading jet p_T (identical to p_T ranges descried in Table 4.2). Lower slices have a generator-level filter, identical to the one in Table 4.3. All numbers are in pb, and each slice is divided into number of additional partons (p) to the $b\bar{b}$ pair (0 to 4)

Slice	$b\bar{b}$ + 0p	$b\bar{b}$ + 1p	$b\bar{b}$ + 2p	$b\bar{b}$ + 3p	$b\bar{b}$ + 4p
J1 (4j17_3j25)	543	588	–	–	–
J2 (4j17_3j25)	283	2.09×10^{3}	7.53×10^{3}	2.89×10^{3}	735
J3 (4j17_3j25)	40.5	1.65×10^{3}	4.93×10^{3}	2.57×10^{3}	971
J4	69.3	456	571	373	229
J5+	1.34	9.04	16.2	14.6	14.1

Background

Most of the backgrounds considered for this channel are of the hadronic nature (mostly jets, arising from partons). Hadronic backgrounds have large cross sections, and thus require many events to reach the same statistics as the data collected. This

Table 4.5 Cross sections and generators for $t\bar{t}$ samples considered for this analysis. Fully hadronic refers to both tops decaying hadronically, semi-leptonic refers to one top decaying hadronically, the other leptonically. Semi-leptonic samples also divided into number of additional partons (p) in the final state

Sample	Generator	σ [pb]
$t\bar{t}\rightarrow$ fully hadronic	MC@NLO + Jimmy	66.48
$t\bar{t}\rightarrow$ semi-leptonic+0p	Alpgen + Jimmy	13.86
$t\bar{t}\rightarrow$ semi-leptonic+1p	Alpgen + Jimmy	13.69
$t\bar{t}\rightarrow$ semi-leptonic+2p	Alpgen + Jimmy	8.47
$t\bar{t}\rightarrow$ semi-leptonic+3p	Alpgen + Jimmy	3.78
$t\bar{t}\rightarrow$ semi-leptonic+4p	Alpgen + Jimmy	1.34
$t\bar{t}\rightarrow$ semi-leptonic+5(+)p	Alpgen + Jimmy	0.50

in turn requires large amounts of computing resources to produce. Many of the backgrounds were similar to those used by the Top Working Group at ATLAS [60], which saved on double production of MC samples.

Two different types of purely hadronic samples were considered: Pythia dijet (as described in Table 4.2), and Alpgen multijet (Table 4.3) and $b\bar{b}$+jets (Table 4.4). These samples are divided into p_T 'slices' (Jn), based off dijet p_T (Pythia) and leading jet p_T (Alpgen). The ranges are described in Table 4.2.

The production and decay of top-antitop ($t\bar{t}$) is also considered a background for this analysis. The top primarily decays to a W boson and a b-quark. The W can then decay either hadronically ($W^+ \rightarrow q^{+2/3}\bar{q}^{+1/3}$) or leptonically ($W^+ \rightarrow l^+\nu_l$). The top pair is then generally divided into three types: fully hadronic, semi leptonic and fully leptonic. For the purposes of this analysis, only the fully hadronic and semi leptonic channels are considered, with samples outlined in Table 4.5.

Chapter 5
Analysis and Trigger Strategy

The VBF $H^0 \rightarrow b\bar{b}$ search channel has not been studied previously in the ATLAS collaboration. Some truth level studies were conducted for this channel for a generic analysis [61], but the work was very preliminary, and did not include any detector specific information. It did, however, list a few features of signal that could be exploited in the event selection: the presence of two "high-p_T" b-jets, showing an invariant mass peak, and a pair of jets in the forward and backward rapidity regions.

Since no previous group at ATLAS had developed an analysis for this channel, a strategy needed to be developed. In addition, an algorithm for the ATLAS trigger system needed to be designed and implemented in order to record the events of interest. Most other search channels used one of the several established lepton triggers in place during data taking, which were centrally maintained and studied.

In this work, trigger and analysis strategies were developed hand-in-hand in order to optimize the identification and selection of signal events. This required the determination of event variables capable of distinguishing between signal and background processes and the development of a new trigger capable of recording signal events while remaining within the limits of the data collection system at ATLAS.

5.1 Event Selection Variables

The general shape of VBF $H^0 \rightarrow b\bar{b}$ events (often referred to as "event topology") resembles the schematic in Fig. 5.1 (recall the Feynman diagram for this channel, shown in Fig. 1.2). Two b-quarks (leading to b-jets) arise from the Higgs decay, and two forward-backward quarks result from the hard scatter (leading to high-p_T jets) and have a large separation in rapidity, η, which leads to a large invariant mass between their jets, m_{jj}.

The event topology of signal events can be utilized to separate between signal and background processes. However, for any given event, the various reconstructed jets first need to be identified as coming from the forward-backward quarks, the Higgs decay, or additional partons. To identify the jets from the forward-backward quarks, a loop over all jet pairs is conducted on signal and background QCD MC

E. Ouellette, *Search for the Higgs Boson in the Vector Boson Fusion Channel at the ATLAS Detector*, DOI 10.1007/978-3-319-13599-1_5

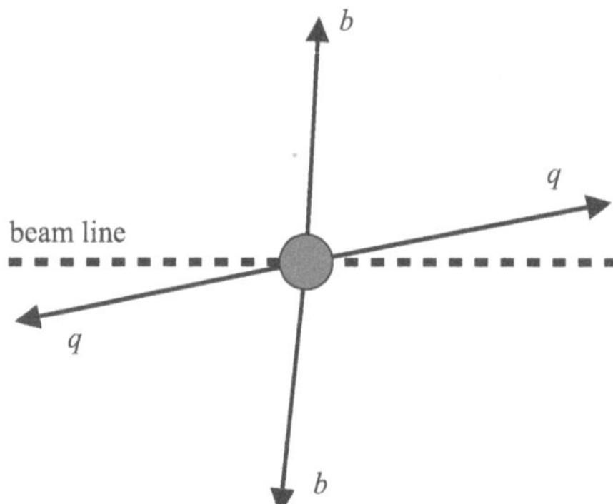

Fig. 5.1 Schematic view of example signal event for VBF $H^0 \to b\bar{b}$ along the beam line

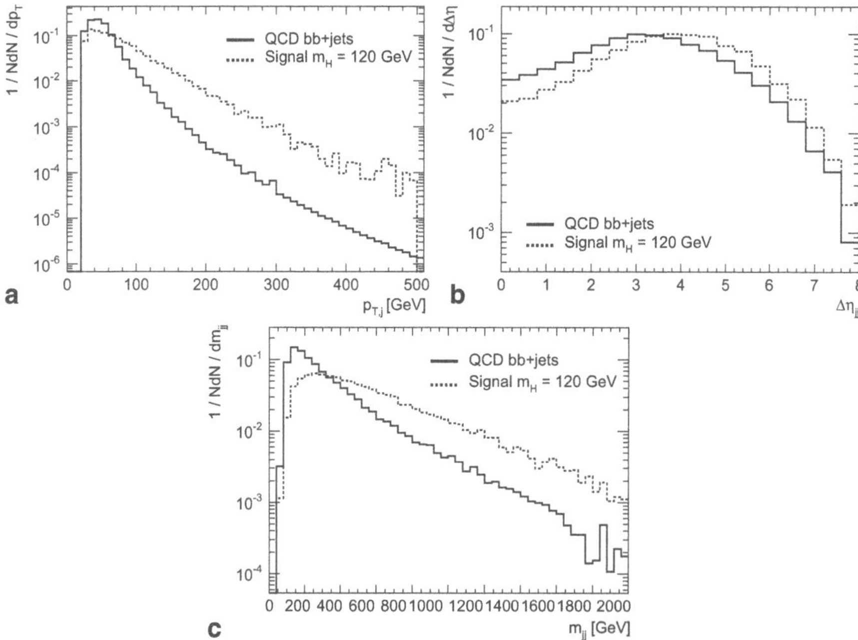

Fig. 5.2 Plots of **a** the p_T of each of the forward-backward jets, **b** $\Delta\eta_{jj}$, their separation in η, and **c** m_{jj}, their invariant mass for both signal and QCD bb+jets MC samples. All plots are normalized to the number of events, and the only selection criterion applied is requiring at least four jets with $p_T > 25$ GeV

samples (from Table 4.4), and the jet pair with the largest invariant mass is taken to be these forward jets. The higher p_T (or "harder") spectrum of signal is evident when comparing the normalized p_T distributions of each of these forward-backward jets for signal and background, plotted in Fig. 5.2a. This harder spectrum, along with the

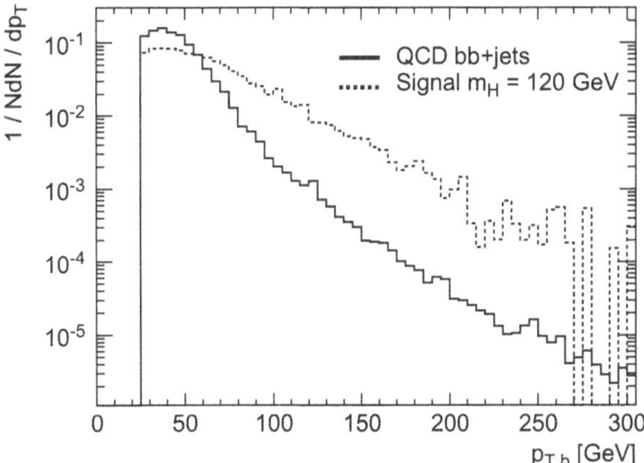

Fig. 5.3 Plots of the p_T of each of the two leading b-jets for signal and QCD bb+jets MC samples. Plots are normalized to the number of events, and the selection criteria applied are requiring at least four jets with $p_T > 25$ GeV, and minimum two b-jets with $|\eta| < 2.5$

larger rapidity gap between these forward-backward jets (plotted in Fig. 5.2b), also leads to a greater m_{jj} distribution (plotted in Fig. 5.2c).

The additional requirement of two b-jets can also enhance the signal. As b-quarks are much less massive compared to the Higgs, the decay products tend to have large momentum. Using the same signal and background samples as above, the p_T distributions of each of the two leading b-jets are plotted in Fig. 5.3, which clearly show signal having a harder p_T spectrum compared to the QCD background.

Another feature of this channel is a suppression of additional radiation in the central region due to a lack of colour exchange between the scattered quarks of the VBF process. In QCD events, the exchange of colour between quarks often leads to additional radiation. Thus, to reduce the contamination from QCD events, a Central Jet Veto (CJV) is used, where events with additional low p_T jets in the central region are vetoed. This is shown in Fig. 5.4, which plots the number of additional jets in the central region (after subtracting the forward-backward and b-tagged jets). Nearly 80 % of signal events have no additional jets, whereas 40 % of QCD events have at least one additional jet in the central region.

5.1.1 Initial Selection Criteria

The event topology was used to develop a set of loose preselection criteria. The selection criteria are optimized using a rank optimization technique. This involves determining the single selection criterion (from a wide set of possible variables) that leads to the greatest significance ($S/\sqrt{S+B}$ was used for this analysis) using MC samples for both signal and background. Once this selection has been determined,

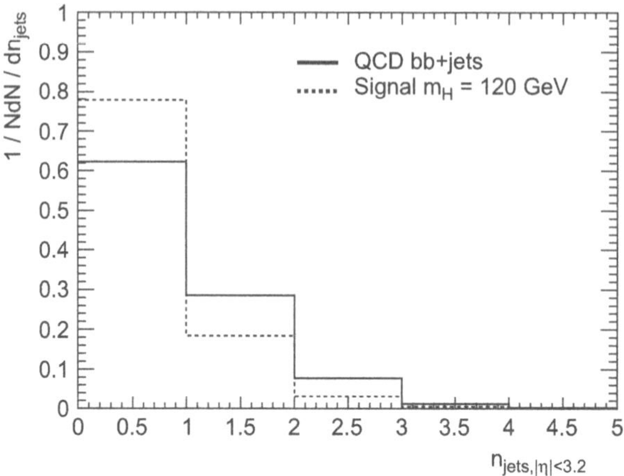

Fig. 5.4 Plots of the number of central jets ($|\eta| < 3.2$) for signal and QCD *bb*+jets MC samples, after subtracting forward-backward- and *b*-jets. Plots are normalized to the number of events, and the selection criteria applied are requiring at least four jets with $p_T > 25$ GeV, minimum two *b*-jets with $|\eta| < 2.5$, and $\Delta\eta_{jj} > 3$

it is applied to the MC samples, and the process is started again: the next selection criterion that maximizes significance is determined, and applied to the dataset. Using the same MC samples as above ($m_H = 120$ GeV signal and QCD background listed in Table 4.4), an early iteration of selection criteria were determined, applied to a range of event variables (mostly the ones mentioned above). This led to $S/\sqrt{S+B} = 0.4$[1] when the MC samples were scaled to 1 fb^{-1}. It is important to stress that these were simply preselection criteria, *without* trigger. Once a proper trigger was chosen, more optimized criteria were determined.

5.2 Early Trigger Development

Any analysis at ATLAS requires a trigger that (a) records signal events efficiently, (b) remains "on" and unprescaled[2] throughout data taking, and (c) ideally reduces background. Using the initial selection criteria, a search for an existing trigger that

[1] Note that the method used to determine the optimized selection criteria used "training" samples (of both MC signal and background samples), however, the resulting significance is calculated using "testing" samples, which is completely different MC samples of both signal and background. This was done to avoid sample bias in the selection criteria.

[2] If a certain trigger has thresholds that are too low, it may record events at a rate above the software and hardware set limits of the trigger system (as detailed in Sect. 4.3.1). Therefore, it often becomes "prescaled", meaning only a fraction of triggered events are actually recorded. When a trigger is "unprescaled", every event that satisfies the trigger requirement is recorded.

satisfies all of these requirements was conducted. Not surprisingly, the most efficient triggers (after selection criteria) where single- and multi-jet triggers with low p_T thresholds, like EF_2j20[3], EF_2j40 and EF_j80. Signal trigger efficiencies on MC samples ranged from 72 to 99 %[4]. However, these triggers were not useful for the analysis, as their thresholds were too low, and data taking rates were far too large. Each of these triggers were either heavily prescaled during the 2011 data taking period, or they were not included, replaced with higher threshold triggers.

The biggest issue for the all-hadronic triggers at L1 was their sensitivity to luminosity, with increases in luminosity resulting in unmanageable rates and dead time for low threshold, low jet-multiplicity triggers. L1_4J10 was ultimately chosen as the L1 trigger for this analysis, as it was determined to have trigger rates that stayed well within the hardware set limits of the trigger system over the entire 2011 data taking period; therefore, it ran unprescaled. In addition, this trigger reduced signal only slightly: the requirement of four jets with p_T thresholds ranging from 30 to 50 GeV[5], which, if applied to the distributions in Figs. 5.2 and 5.3, still left significant signal.

The trigger efficiency for L1_4J10 was 17 % when using the signal MC samples. Although this trigger did not cut on signal too tightly, it was only the L1 seed (as mentioned in Sect. 4.3.1, events triggered at L1 are "seeds" for triggers at HLT), and further requirements were added at HLT for it to stay within the software set limits at this higher trigger level. There was a desire that these additional trigger requirements would (a) reduce background in order to stay within the trigger rate limits and (b) be applied offline as well, such that the effect on signal would be minimal.

5.2.1 *b*-jet Trigger

The next logical idea was to implement a b-jet trigger at HLT. Early trigger menus in ATLAS had rudimentary b-jet triggers; however, great progress and improvements were made in 2011 to the b-jet trigger algorithm. Similar to the b-jet algorithm described in Sect. 4.4.2, the online b-tagging algorithm uses inner detector information to identify jets originating from b-quark decays. No tracking information is available at L1, thus b-tagging is only possible at HLT.

The online b-tagging algorithm [62] is a simplified version of the offline b-tagging algorithm, utilizing the impact parameter, d_0, as a discriminating variable, which is

---

[3] Note that trigger signatures used at ATLAS are represented by a character string that always begins with the level of trigger considered (L1, L2 or EF), followed by strings that describe the cuts applied. For example EF_*nj*X, where *n* represents the number of jets, and *X* describes the p_T threshold. If no *n* is printed, a single jet trigger is implied.

[4] Trigger efficiencies are the ratio of the number of events that satisfy a trigger and the total number of events.

[5] The L1 jet trigger thresholds (eg. J10) does not directly represent the p_T of the reconstructed jet, but rather a very rudimentary approximation of it. In reality, the J10 requirement at L1 translates to an offline jet with $p_T > 30$ GeV typically.

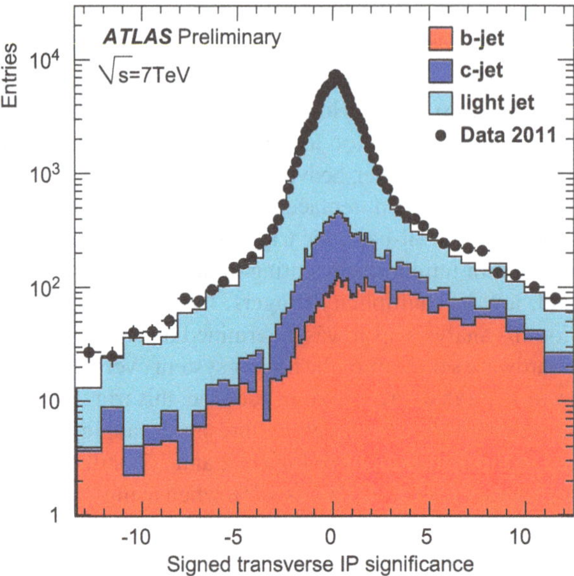

Fig. 5.5 Data and MC samples showing the signed impact parameter significance of jets. This variable is used in the online b-tagging algorithm, since b-jets tend to have a more positive impact parameter significance than light- and c-jets. Taken from [62]

the distance of closest approach between a particle track and the primary vertex. In the HLT, tracks in the inner detector are associated with L1 jets. The impact parameters are combined with their uncertainty, $\sigma(d_0)$, to form the impact parameter significance, $S(d_0) = d_0/\sigma(d_0)$, the primary discriminant of the online b-tagging algorithm. The b-jets tend to have a more positive impact parameter significance compared to light jets, as seen in Fig. 5.5. A probability is derived for each track associated to a jet, and combined together to form a per jet probability, measuring a likelihood to be a b-jet. Online selection criteria are then made to require that n-jets satisfy either a loose, medium or tight selection, where a loose (tight) selection has high (low) b-tagging efficiency.

5.3 2011 Trigger Menu

The implementation of a dedicated b-jet trigger for VBF $H^0 \rightarrow b\bar{b}$ into the ATLAS trigger menu required working closely with trigger experts, since they had to ensure a new item remained within the hardware and software bounds of the detector. However, the new trigger had to select sufficient signal in order for this analysis to be possible. In 2011, after L1_4J10 was approved as an unprescaled L1 trigger seed, several different options were considered at HLT, using various combinations of number of b-jets, and loose, medium or tight selection on those jets.

The considered triggers all had similar trigger efficiencies; however, approximate data taking rates (for a simulated luminosity of 1×10^{33} cm^{-2}s^{-1}) varied from one trigger to another, with trigger rates ranging from 4 to 25 Hz. A rate above 10 Hz was considered far too high, considering the total allowed rate (at EF) over all trigger items was $\sim$200 Hz (as mentioned in Sect. 4.3.1). Therefore, the trigger with the lowest rate, EF_2b10_medium_4L1J10[6], was chosen.

5.3.1 Addition of HLT Jets

The chosen trigger, EF_2b10_medium_4L1J10, was approved up to a luminosity of 2×10^{33} cm^{-2}s^{-1}. As the plot of peak luminosity as a function of date in Fig. 4.6 shows, that luminosity was reached between July and August 2011, and the trigger was turned off. Fortunately, the L1 was not an issue, since it had been approved up to 5×10^{33} cm^{-2}s^{-1}, since its estimated rate of 2.7 kHz was small compared to the total allowed L1 rate of 65 to 70 kHz.

A natural extension was to require HLT jets in the trigger; the previous trigger only cut on the p_T of jets at L1, whereas only b-tagging was used at HLT. The additional requirement of four online jets that satisfy $p_T \gtrsim 30$ GeV was tested. Although this reduced the signal trigger efficiency, it reduced background by a greater factor, and lowered the data taking rate by an order of magnitude. The resulting trigger implemented within the menu was EF_2b10_medium_4j30_a4tc_EFFS[7].

5.4 Trigger MC Correction

It is difficult to properly simulate all detector effects when producing MC samples. One notoriously difficult quantity to reproduce is the online and offline b-tagging of jets. Therefore, scale factors are calculated from data and the MC samples to account for these differences. Since this analysis uses both online and offline b-tagging, an adapted b-tagging correction [63] is applied by calculating an event-by-event scale factor, derived from per jet weights. For example, if a certain jet is tagged both online AND offline (referred to as a "fat tagged" jet), the corresponding jet weight, w_{jet}, is:

$$w_{\text{jet}} = \text{SF}_f(p_T, \eta) = \frac{\varepsilon_f^{\text{Data}}(p_T, \eta)}{\varepsilon_f^{\text{MC}}(p_T, \eta)} \tag{5.1}$$

[6] The trigger EF_2b10_medium_4L1J10 required L1_4J10 to be used as a seed (4L1J10), and two b-jets (2b10) with "medium" selection criteria (medium).

[7] The trigger EF_2b10_medium_4j30_a4tc_EFFS satisfies the same requirements as EF_2b10_medium_4L1J10, and the requirement of four jets with $p_T \gtrsim 30$ GeV (4j30), reconstructed using the anti-k_T algorithm with $C = 0.4$ (see Eq. 4.3) and topoclusters as objects (a4tc), with a full scan at EF (EFFS).

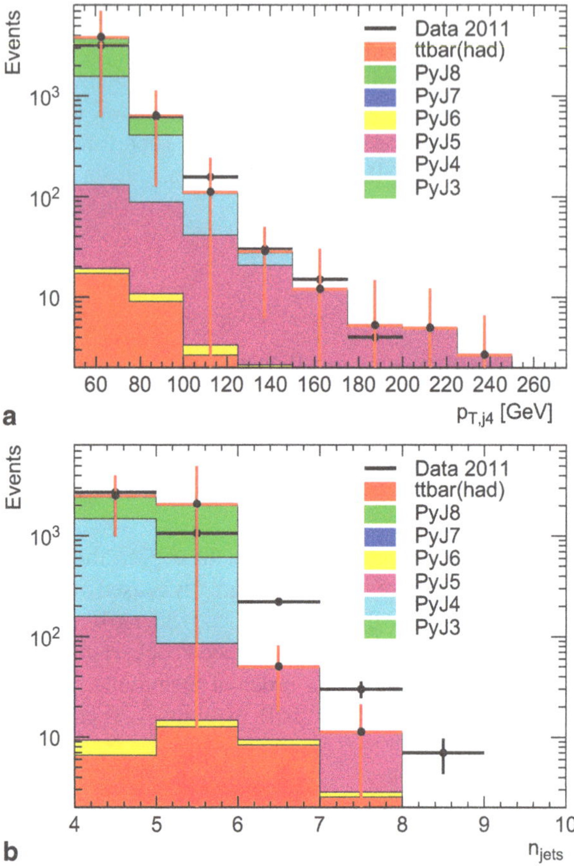

Fig. 5.6 Plots of **a** fourth leading jet p_T, and **b** number of jets above 50 GeV, with both MC samples and data after selection criteria have been applied. Note that the data corresponds to 0.8 fb^{-1} of 2011 data, and the red lines show the statistical uncertainty on the MC samples

where f is the flavour (light, or b[8]) of the simulated jet, $\varepsilon_f^{\text{Data}}$ ($\varepsilon_f^{\text{MC}}$) is the efficiency in data (MC) that a jet of flavour f is fat tagged[9], depending on the jet's p_T and η. Similarly, if the jet is not fat tagged, the per jet weight is:

$$w_{\text{jet}} = \frac{1 - \varepsilon_f^{\text{Data}}(p_T, \eta)}{1 - \varepsilon_f^{\text{MC}}(p_T, \eta)} = \frac{1 - \text{SF}_f(p_T, \eta)\varepsilon_f^{\text{MC}}(p_T, \eta)}{1 - \varepsilon_f^{\text{MC}}(p_T, \eta)}. \tag{5.2}$$

[8] The c-jets have the same scale factors as b-jets due to low statistics in the scale factor measurements.

[9] Efficiencies for MC samples are determined by using the flavour of jets taken from truth level information. Efficiencies for data are determined by extracting highly purified samples of b-jets, using various methods.

An event weight is computed as the product of all individual jet weights in the event:

$$w_{\text{event}} = \prod_{N_{\text{jets}}} w_{\text{jet}}, \tag{5.3}$$

which is applied as a scale factor to MC samples when comparing them to data.

This was done for early 2011 data, and Pythia jet-jet MC samples. Unfortunately, the statistics of the MC samples were the limiting factor, since not nearly enough MC samples were available to account for the 0.8 fb^{-1} of data used for the comparisons, shown in Fig. 5.6. There does appear to be approximate agreement, though the low statistics suggested these MC samples could not be used to estimate the background for this channel, as statistical uncertainties would dominate the search for this signal in the data set.

Chapter 6
Final Analysis Strategy

Following the implementation of the triggers mentioned in Sect. 5.3, it was possible to finalize the analysis strategy for the VBF $H^0 \rightarrow b\bar{b}$ search channel.

6.1 Trigger Selection

As mentioned in Chap. 5, a dedicated trigger was developed for this channel[1], with L1_4J10 used as a trigger seed at L1, as this was the lowest unprescaled multijet trigger during 2011 data taking. The requirement of two online b-tagged jets were added at HLT (EF_2b10_medium_4L1J10) and was approved up to an instantaneous luminosity of 2×10^{33} cm^{-2} s^{-1}. Once this luminosity was reached, HLT jets were added to the trigger requirements (EF_2b10_medium_4j30_a4tc_EFFS). These triggers cut quite tightly on the signal MC samples, as they have trigger efficiencies of 6.2 % and 2.3 % for EF_2b10_medium_4L1J10 and EF_2b10_medium_4j30_a4tc_EFFS, respectively.

The $\sqrt{s} = 7$ TeV data used for this analysis was collected by ATLAS during the 2011 proton-proton run at the LHC and amounts to 4.2 fb^{-1} of integrated luminosity: 1.25 fb^{-1} collected using EF_2b10_medium_4L1J10 and 2.94 fb^{-1} using EF_2b10_medium_4j30_a4tc_EFFS.

6.2 Object Selection

Jets are reconstructed from energy clusters in the calorimeters using the anti-k_T jet algorithm using a radius parameter of $C = 0.4$ (as defined in Eq. 4.3). Jet energies are calibrated using p_T- and η-dependent scale factors derived from MC samples and validated using data, as described in Sect. 4.4.1. Background jets can

[1] The trigger was developed by a b-trigger working group within ATLAS, in consultation with myself.

E. Ouellette, *Search for the Higgs Boson in the Vector Boson Fusion Channel at the ATLAS Detector*, DOI 10.1007/978-3-319-13599-1_6

be produced by beam-gas or beam-halo events, cosmic rays passing through the detector, or calorimeter noise. To ensure a high rejection of background jets, only jets that satisfy a set of quality selection criteria are used. This selection ranges from cuts on the energy fraction in the EM and hadronic calorimeters, the quality of the LAr pulse shapes of the cells in a jet, timing, and other criteria. The selection used for this analysis has a 99.8 % efficiency in selecting good jets [64].

For any given event, there can be several collisions occurring at once, which leads to pile-up events occurring in the same bunch crossing. To reduce the jets coming from these pile-up events, jets are required to have at least 75 % of the transverse momentum of their associated inner detector tracks above 0.4 GeV to originate from the primary vertex. Furthermore, to reduce fake jets originating from electrons, any jet that lies within $\Delta R < 0.2$ of an identified electron is removed. Finally, only jets with $|\eta| < 4.5$ and $p_T > 25$ GeV are considered.

Jets originating from b-quarks are identified using the 57 % b-tag efficiency working point of the JetFitterCOMBNN b-tagging algorithm, as described in Sect. 4.4.2.

6.3 Event Selection

At least four jets are required to have $p_T > 50$ GeV in order for the trigger (L1_4J10) to be fully efficient [65]. As previously mentioned (in Sect. 5.1), the scattered quarks of the VBF process are both hard, and have a large separation in η. This leads to a large invariant mass of the resulting jets, m_{jj}, and difference in η between the jets, $\Delta\eta_{jj}$. The jets most likely to come from these scattered quarks are identified by finding the jet pair (for jets with $p_T > 50$ GeV) that have the largest invariant mass. Selection criteria of $m_{jj} > 584$ GeV and $\Delta\eta_{jj} > 3$ were determined to maximize significance when running a rank optimization on signal and background MC samples.

Of the remaining jets with $p_T > 50$ GeV, those that lie within $|\eta| < 2.5$ are Higgs candidate jets (as they may have come from the Higgs decay to $b\bar{b}$). This spatial requirement ensures that if these jets are also b-tagged, they are within the limits of the tracking system, an essential component of b-tagging algorithms. Of the Higgs candidate jets, one or two of them must also be b-tagged. Any two Higgs candidate jets must also be well separated, and have $\Delta R_{bb} > 0.7$. Since this analysis concentrates on the Higgs mass range of 115 to 130 GeV, only Higgs candidate jet pairs with an invariant mass that satisfies $80 < m_{bb} < 300$ GeV are considered[2].

Little additional gluon radiation is expected in the central region of the detector due to the colourless exchange between the forward-backward scattered quarks,

[2] Note that the subscript bb in ΔR_{bb} and m_{bb} will represent the Higgs candidate jet pair. This will be used to distinguish them from the forward-backward jets pair, which uses the subscript jj. However, the bb pair does not necessarily ensure they are both b-tagged, as you will see below.

especially compared to QCD events. A Central Jet Veto (CJV) is thus applied, which vetoes any event with additional jets that satisfy $|\eta| < 3.2$ and $25 < p_T < 50$ GeV.

6.4 Background Estimation

As the cross sections in Tables 4.2, 4.3 and 4.4 suggest, the background is quite large for this channel. They are so large that the time needed to generate and fully simulate the background MC samples needed to match the amount expected for 4.2 fb^{-1} of integrated luminosity would have required millions of CPU hours. This was deemed too much to request of the ATLAS MC groups. Thus a data driven method was needed to estimate the background.

A method developed specifically for this analysis, based loosely on the one used by CDF [66], estimates the background by calculating the rate at which a jet is b-tagged given that there is already a b-tagged jet in the event. In order to reduce the effect of possible signal contamination, this is done in a background-dominated region of phase space.

The data is divided into three distinct regions: signal, tag and control. Signal is characterized by events with exactly four jets with $p_T > 50$ GeV (4J), tag by events with exactly five jets with $p_T > 50$ GeV (5J), and control by events with six or more jets with $p_T > 50$ GeV (6J+). Events are first selected as per the triggers described in Sect. 6.1 and the selection criteria described in Sect. 6.3. In the tag region, a Tagging Rate Function (TRF) is derived by calculating the ratio between events where exactly two Higgs candidate jets are b-tagged and events where exactly one of a pair of Higgs candidate jets is b-tagged. This is done as a function of the spatial separation between the jet pair (ΔR_{bb}) and the minimum p_T of the two jets:

$$TRF_{\text{tag}}\left(\Delta R_{ij}, min[p_{T,i}, p_{T,j}]\right) = \frac{N_{2b}\left(\Delta R_{ij}, min[p_{T,i}, p_{T,j}]\right)}{N_{1b}\left(\Delta R_{ij}, min[p_{T,i}, p_{T,j}]\right)}. \tag{6.1}$$

The number and width of bins are chosen in such a way to have roughly the same statistics over the whole range of ΔR and p_T. The resulting TRF is plotted in Fig. 6.1.

Once this TRF is calculated, data events are selected in the signal region with the same initial selection criteria as those in the tag region. Events with one-and-only-one b-tagged Higgs candidate jet are used to predict the shape of doubly b-tagged events, by using an event weight taken from the TRF, depending on the ΔR and $min[p_{T,1}, p_{T,2}]$:

$$N_{2b}^{predicted} = N_{1b}^{data}\left(\Delta R_{ij}, min[p_{T,i}, p_{T,j}]\right) \times TRF_{\text{tag}}\left(\Delta R_{ij}, min[p_{T,i}, p_{T,j}]\right) \tag{6.2}$$

The result of this procedure is an estimate of the shape and acceptance of various kinematic variables in events with two b-tagged jets, derived from events with only one b-tagged jet.

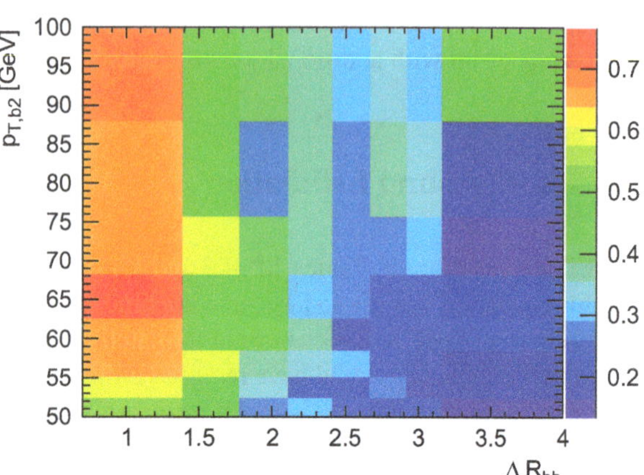

Fig. 6.1 The tag rate function shown as a function of ΔR_{bb} and $p_{T,b2}$, derived from the tag region (5J) on 4.2 fb^{-1} of data

6.4.1 Background Estimation Check

As a check, the method above is reapplied to events in the tag region. The resulting distributions of some standard kinematic variables are plotted in Fig. 6.2 (p_T of leading and sub-leading b-tagged jet, ΔR_{bb}, m_{bb}, and m_{jj} and $\Delta\eta_{jj}$ of the two forward jets). There is very good agreement over all the variables, with the exception of $\Delta\eta_{jj}$, which appears to be systematically shifted. This is believed to be caused by the size and width of b-jets in the event. Jets arising from b-quarks tend to have a greater width, due to the heavy mass of the b-hadron that is created during its decay through the detector. This in turn causes the jets in the event to become more spherical[3] and isotropic, increasing slightly the distance between them. In the plot of $\Delta\eta_{jj}$, the distribution with two b-tagged jets tends to have a greater separation, as it has more b-tagged jets compared to the distribution with a single b-tagged jet. To correct for this, a linear scaling is calculated with respect to $\Delta\eta_{jj}$ and reapplied to the data. The kinematic plots, plotted in Fig. 6.3, show better agreement (all within 10 % agreement). To quantify this improvement, χ^2/NDF was calculated between data and the estimation, both before and after the correction was applied. Their values before the correction ranged from 0.4 to 1.0 for most distributions, except $\Delta\eta_{jj}$, which had $\chi^2/\text{NDF} = 4.0$. This was improved to $\chi^2/\text{NDF} = 0.9$ with the correction applied (all other values remained mostly unchanged).

As an additional check, the TRF was tested in a sideband of the signal region, where $300 < m_{bb} < 400$ GeV. However, statistics are not as good in this region. The resulting kinematic distributions are plotted in Fig. 6.4, with the background estimation and data points still appearing to be roughly in agreement with each other. The χ^2/NDF for each distribution was calculated before and after the $\Delta\eta_{jj}$

[3] Sphericity is a measure of how isotropic the jets in an event are, and is calculated from the eigenvalues of the sphericity tensor, described in [54].

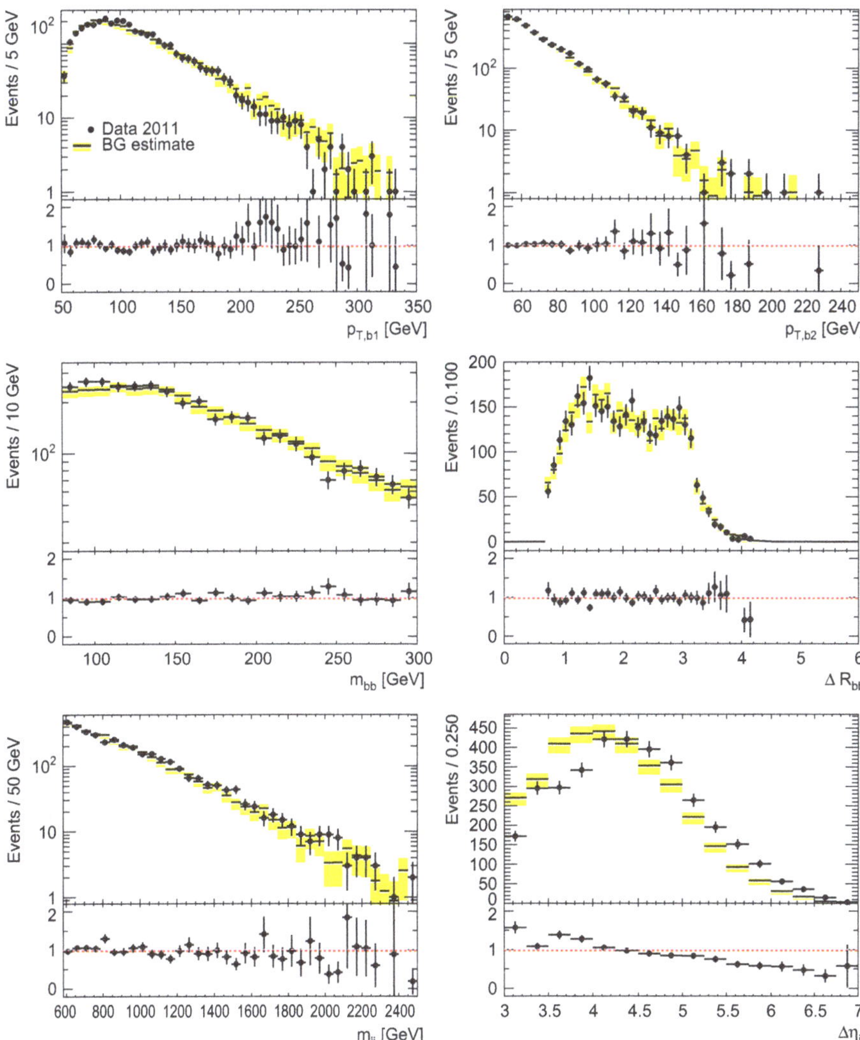

Fig. 6.2 Result of reapplying tag rate function to events in the tag region (5J) over several token kinematic variables. The TRF derived background estimation is shown with statistical error bars in yellow, while the 2b events are the black data points

correction was applied, and similar results were observed: values unchanged (in the range of 0.6 to 1.0) for most distributions, except $\Delta\eta_{jj}$ (improved from 2.7 to 1.2).

Finally, the TRF was also applied to the control region (6J+). The resulting plots, shown in Fig. 6.5, still show agreement between data and the estimation. Similar results were obtained as above when comparing the χ^2/NDF of distributions before and after a $\Delta\eta_{jj}$ scaling was applied.

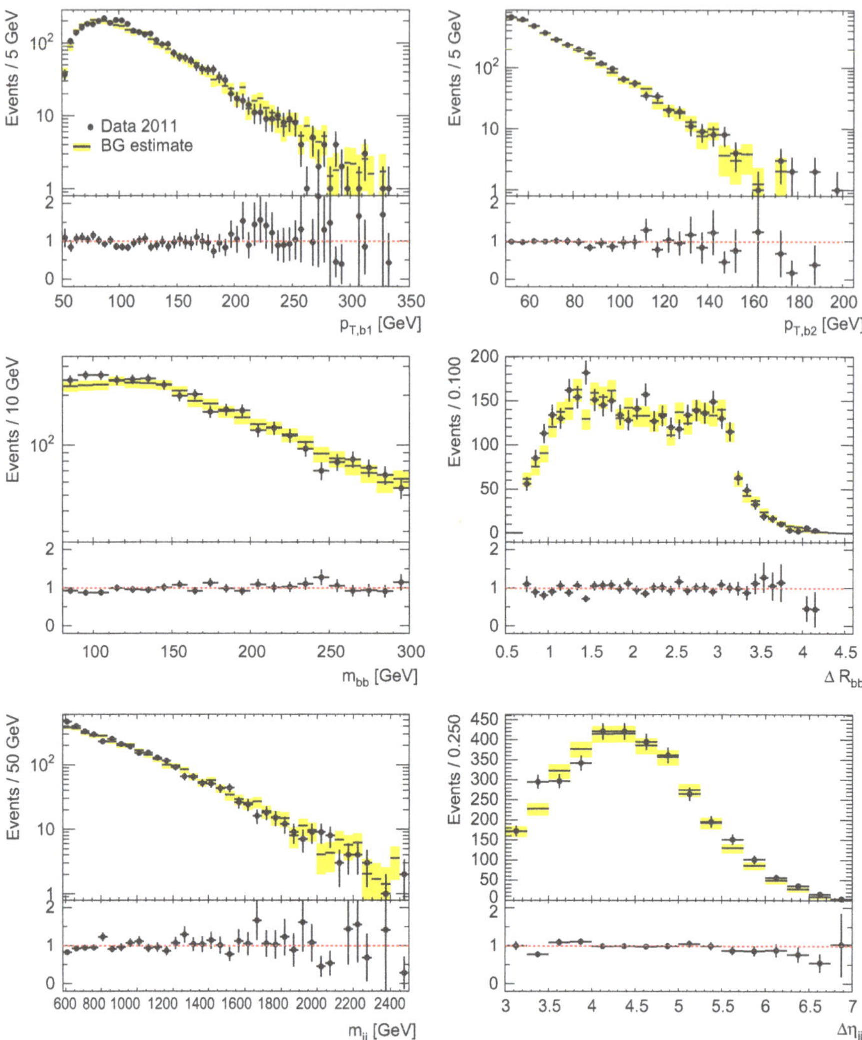

Fig. 6.3 Result of reapplying tag rate function to events in the tag region (5J) over several token kinematic variables, with an additional linear scaling applied with respect to $\Delta\eta_{jj}$. The TRF derived background estimation is shown with statistical error bars in yellow, while the 2b events are the black data points

6.4.2 Background Estimation Results

The application of the tag rate function to singly b-tagged events in the signal (4J) region is shown in Fig. 6.6. These plots show great agreement between the background

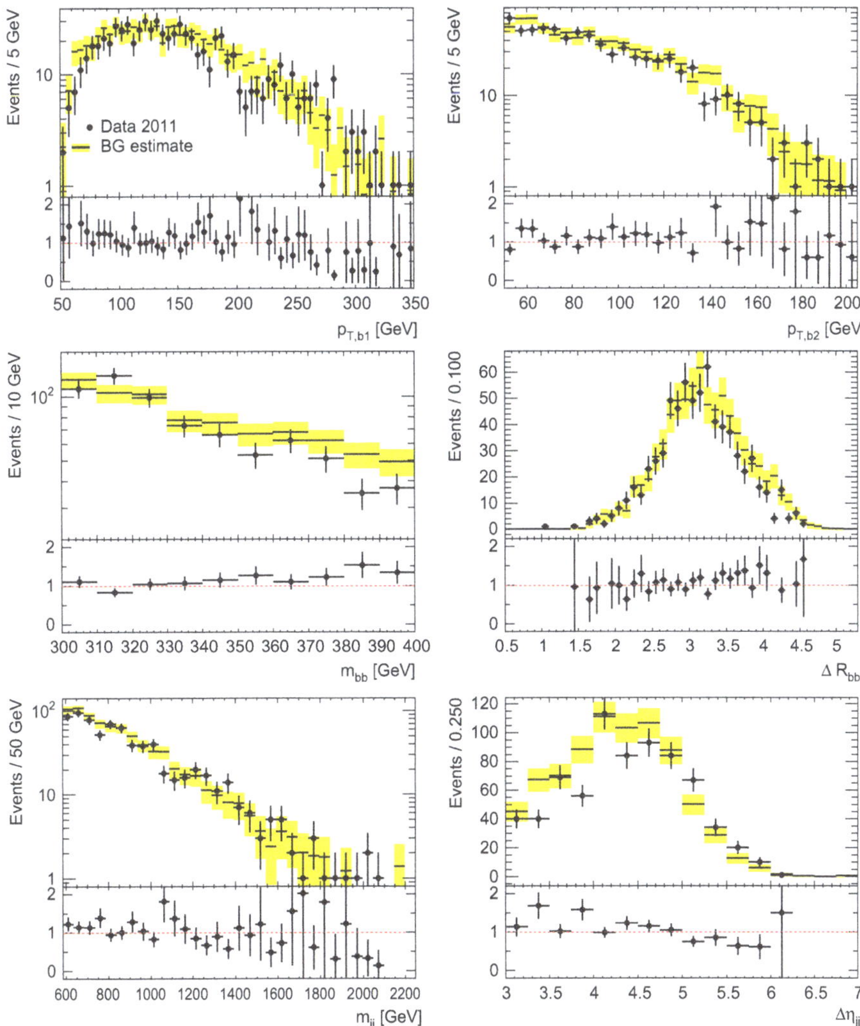

Fig. 6.4 Result of applying tag rate function to events in the signal region (4J) over several token kinematic variables, in a region of no signal (300< m_{bb} < 400 GeV), with an additional linear scaling applied with respect to $\Delta\eta_{jj}$. The TRF derived background estimation is shown with statistical error bars in yellow, while the 2b events are the black data points

estimation and the 2b data sample[4]. The plot of m_{bb} is of particular importance, as this variable is used to discriminate between signal and background. The comparison between data and the background estimation is shown in Fig. 6.7.

[4] Note that although these plots are shown before any analysis results have been given, the result of the background estimation in the signal region was left blinded throughout the analysis to avoid bias.

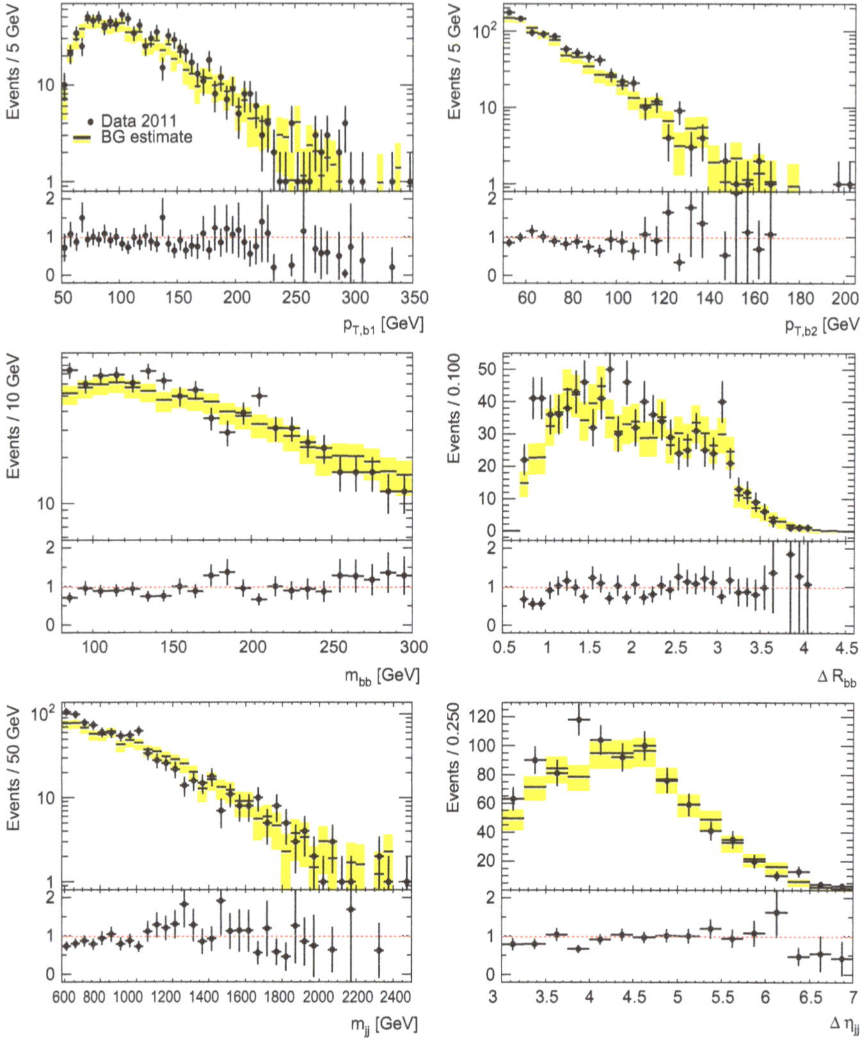

Fig. 6.5 Result of applying tag rate function to events in the control region (6J+) over several token kinematic variables, with an additional linear scaling applied with respect to $\Delta\eta_{jj}$. The TRF derived background estimation is shown with statistical error bars in yellow, while the 2b events are the black data points

6.4.3 Monte Carlo Check

The goal of this background estimation is to reliably estimate the shape and acceptance of QCD background for this channel using a fully data driven technique. This must be verified in some way on actual QCD samples. As mentioned above, the computing resources necessary to produce and simulate sufficient background MC

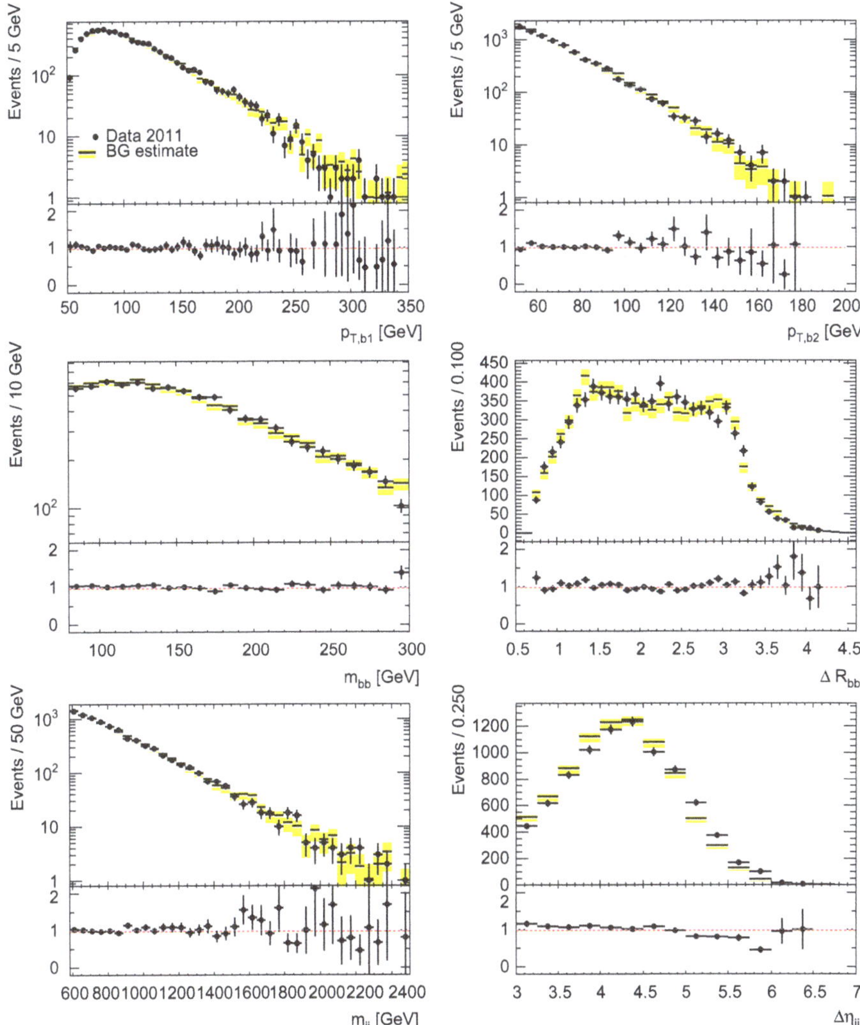

Fig. 6.6 Result of applying tag rate function to events in the signal region (4J) over several token kinematic variables, with an additional linear scaling applied with respect to $\Delta\eta_{jj}$. The TRF derived background estimation is shown with statistical error bars in yellow, while the 2b events are the black data points

samples are far too great for proper comparison. There exist small samples of fully simulated QCD multijet and $b\bar{b}$+jets samples (as outlined in Tables 4.3 and 4.4), but when the background estimation was run on these samples as a check of their reliability, the statistics were far too low to glean any meaningful information from the results.

Instead, a large statistic sample of Pythia QCD background events was produced, with four-vectors of the jet products being used to reconstruct truth level jets, but

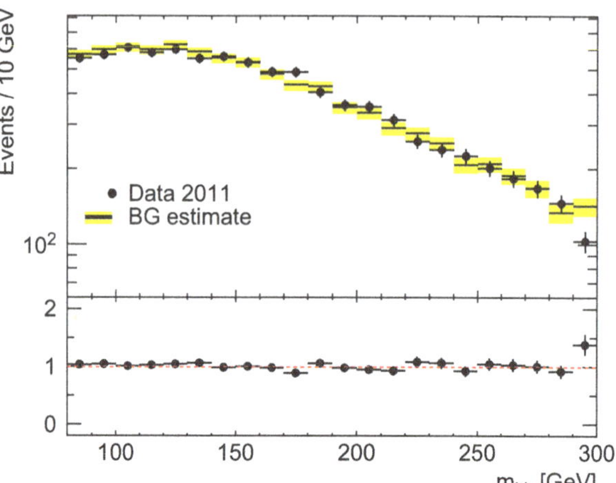

Fig. 6.7 Plot of the m_{bb} distribution taken from 2b (Data 2011) and 1b (BG estimate) samples with two b-tagged jets in the signal (4J) region

without simulating the passage of the particles through the detector. This was used to run a truth level study of this analysis, and test the reliability of this method at estimating background.

No detector information was available, thus it was impossible to apply an appropriate trigger or any jet b-tagging. It was possible, however, to determine the quark flavour of each truth level jet (light, charm or bottom); this was used in lieu of b-tagging. The methodology outlined above was tested on this sample: the application of the same selection criteria as Sect. 6.3, a TRF was derived in the tag region, and the resulting TRF was reapplied to tag, control and signal regions. Similar to the data samples, a shift in $\Delta\eta_{jj}$ was seen, once again caused by the increased sphericity in the event with added b-jets in the 2b case compared to the 1b case. This was corrected using a linear scaling, just like the data sample. After application of the TRF and the linear correction, the estimates were successfully able to match the signal. To quantify this improvement, the signal and the estimation distributions were compared using χ^2/NDF (just as in data). Although the high statistic nature of this sample lead to higher values of χ^2/NDF (compared to values calculated in Sect. 6.4.1), the application of a linear scaling improved χ^2/NDF for the $\Delta\eta_{jj}$ distribution significantly (from 39 before to 5 after). The resulting kinematic plots are shown in Fig. 6.8 in the 4J region, with a $\Delta\eta_{jj}$ linear correction applied. In general the single b-jet background estimation matched the double b-jet distributions within 15 %. To compensate for these small discrepancies, a shape systematic is derived from the difference between the 1b estimation and the 2b distributions, and applied to the background estimation for data. This will be outlined further in the next chapter.

6.4.4 Influence from Other Background Sources

This method is used primarily to estimate the backgrounds from QCD events. However, other non-QCD sources may contribute to the background of this channel,

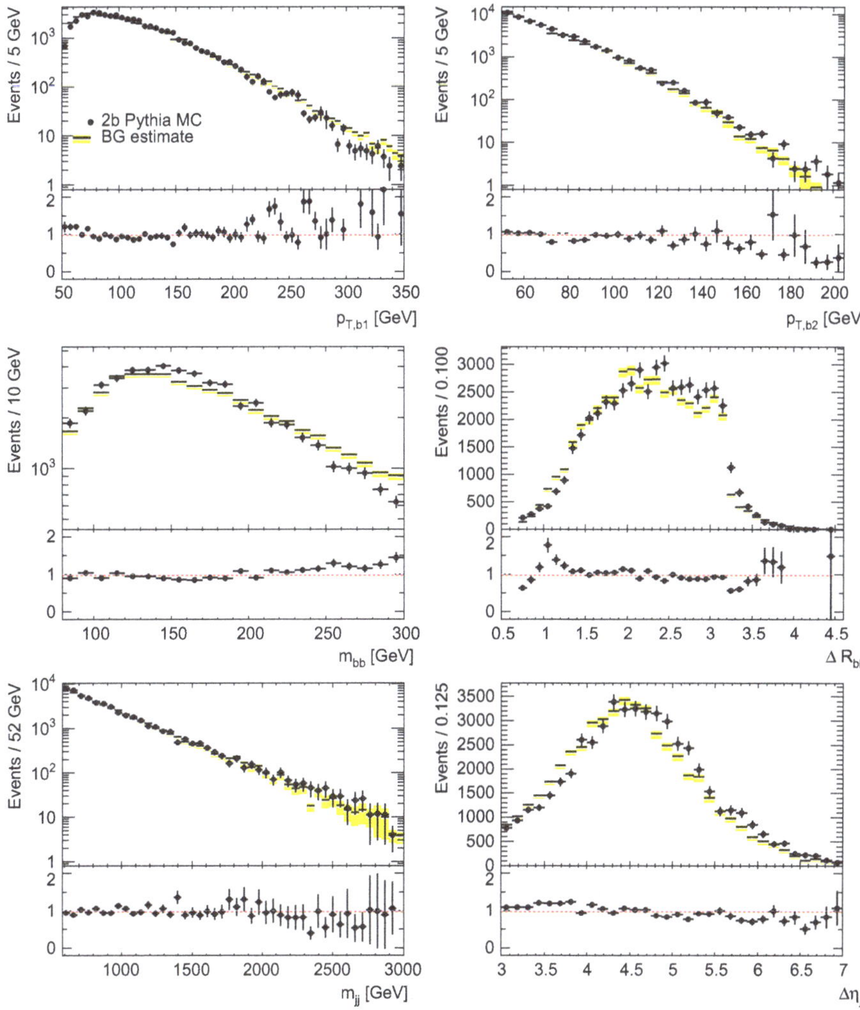

Fig. 6.8 Result of running a MC check of the TRF method on a Pythia sample. The above plots show standard kinematic variables in the signal (4J) region of 2 b-jet events, compared with the TRF derived estimate of from 1 b-jet events. Note that this is also after a $\Delta\eta_{jj}$ scaling has been applied

but not behave the same as QCD when put through the background estimation. If MC samples of these sources were used as a separate background in the analysis, double counting would occur when applying the TRF to singly b-tagged events on the same sample. Therefore, the background estimation must have the 1b non-QCD contributions removed in order to avoid this, since:

$$\begin{aligned}\text{Background estimation} &= 1b\ \text{Data} \times TRF\\ &= [1b\ QCD + 1b\ \text{non-QCD}] \times TRF,\end{aligned}$$

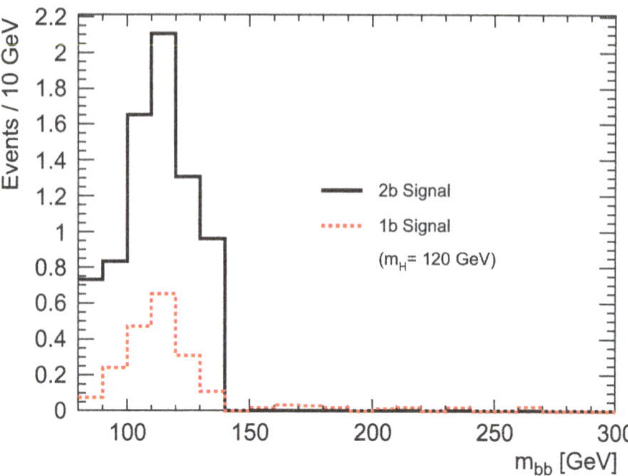

Fig. 6.9 Plot of the m_{bb} distribution taken from the 2b and 1b samples in the signal (4J) region using the $m_H = 120$ GeV MC signal samples, scaled to 4.2 fb^{-1}

therefore to get an estimate of the QCD background only, one must subtract the 1b non-QCD contributions from the TRF derived estimation:

$$\text{QCD estimation} = (1b \text{ Data} \times TRF) - (1b \text{ non-QCD} \times TRF). \tag{6.3}$$

The largest source of background after QCD is $t\bar{t}$, since it has a high jet multiplicity, plus can have several b-jets. For this analysis, only fully-hadronic and semi-leptonic $t\bar{t}$ decays are considered, with MC samples listed in Table 4.5.

Other backgrounds were considered, such as $W \rightarrow$ hadronic+jets, $Z \rightarrow b\bar{b}$+jets, and WW, WZ and ZZ diboson samples. Very few events survived the event selection, due mainly to the selection criteria on m_{jj} and $\Delta\eta_{jj}$. Their effect on the analysis is negligible.

6.4.5 *Influence from a Possible Higgs signal*

If a Higgs signal existed, it would also contribute to the background estimation, just as the other sources of MC background samples contribute above. Even though the Higgs signal is expected to be many times smaller than the background, its cross section is inflated when performing an exclusion limit calculation (described in Appendix C, and further explained in Chap. 8). Therefore, just as the non-QCD source were removed from background, the 1b Higgs signal contributions, scaled by a factor of n, are removed as well:

$$\begin{aligned}\text{Background estimation} &= 1b \text{ Data} \times TRF \\ &= [1b\ QCD + 1b \text{ non-QCD} + n \times (1b \text{ Higgs})] \times TRF,\end{aligned}$$

therefore to get the overall background estimation, one subtracts the other 1b contributions:

$$\text{QCD estimation} = (1b \text{ Data} \times TRF)$$

$$- (1b \text{ non-QCD} \times TRF)$$
$$- n \times (1b \text{ Higgs} \times TRF).$$

On average, the 1b contribution from signal MC samples is roughly 30 % that of the total 2b distribution in the signal region (4J), as shown in Fig. 6.9.

Chapter 7
Systematic Uncertainties

The 95 % CL exclusion calculation, as outlined in Appendix C, requires an understanding of the *nuisance parameters* of both signal and background, in order to set limits on the signal strength, μ. The nuisance parameters are more commonly known as systematic uncertainties. These are described in detail here.

7.1 Background Estimation Systematic Uncertainties

The method used to estimate the background has its limitations. It is important to quantify the several sources that affect both the shape and acceptance of the estimation plotted in Fig. 6.7. These are summarized in Table 7.1.

A natural bias occurs from deriving the TRF using the tag region (5J). To account for this, an alternate TRF was derived from a control region (6J+). In general, most individual bin values (like those in Fig. 6.1) for the control derived TRF are within ± 20 % of the same bin value for the tag derived TRF. However, some reach has high as $+90$ % and as low as -70 %. The resulting background estimation when applying this TRF to the signal region is used as a shape systematic. Plots of the difference in m_{bb} using either TRF are shown in Fig. 7.1. Despite the fluctuation in the TRF, the variations in the m_{bb} only range from $+8$ to -8 %.

The linear correction applied to the estimation as a function of $\Delta\eta_{jj}$ also incurs a bias on the shape of the background, as it is derived from the difference between the $\Delta\eta_{jj}$ plots for data and the background estimation in the tag region (5J). To account for this bias, an alternate correction is derived from the $\Delta\eta_{jj}$ plots of data and background estimation in the control region (6J+). This new linear correction is then applied to events in the signal region. A plot of the resulting m_{bb} distributions using both corrections is shown in Fig. 7.2. The variation per bin ranges from $+3$ to $+4$ %.

Finally, as mentioned in Sect. 6.4.3, a systematic uncertainty has been assigned to the background estimation due to the small variations found when running a MC check on the background estimation method. An average relative uncertainty (per-bin) was derived from the m_{bb} distribution in Fig. 6.8, with the result plotted in Fig. 7.3a. As one can see from the this plot, relative uncertainties range from a

E. Ouellette, *Search for the Higgs Boson in the Vector Boson Fusion Channel at the ATLAS Detector*, DOI 10.1007/978-3-319-13599-1_7

Table 7.1 Summary of the sources of uncertainty on the background estimation, and their overall effect on acceptance

Source of uncertainty	Effect on background estimation (%)
TRF derivation	+5
$\Delta\eta_{jj}$ scaling	+3
MC uncertainty	±10

minimum of 6 % to a maximum of 20 %. This relative uncertainty is then propagated to the actual background estimation distribution of m_{bb}, with the result plotted in in Fig. 7.3b. This is a conservative estimate of the uncertainty in the background estimation method.

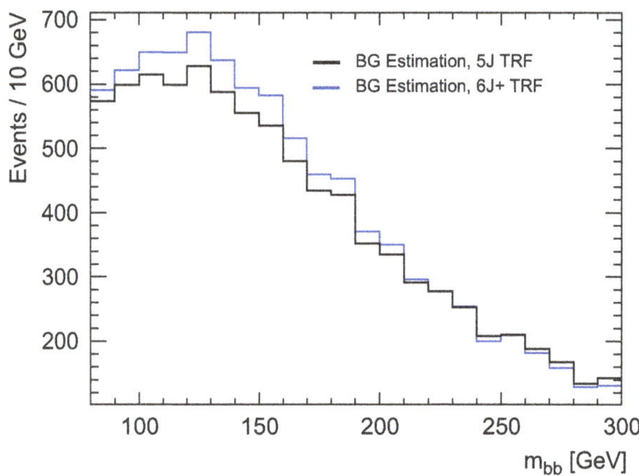

Fig. 7.1 Plots of m_{bb} using the background estimation derived from the tag region TRF (*black*) and the control region TRF (*blue*)

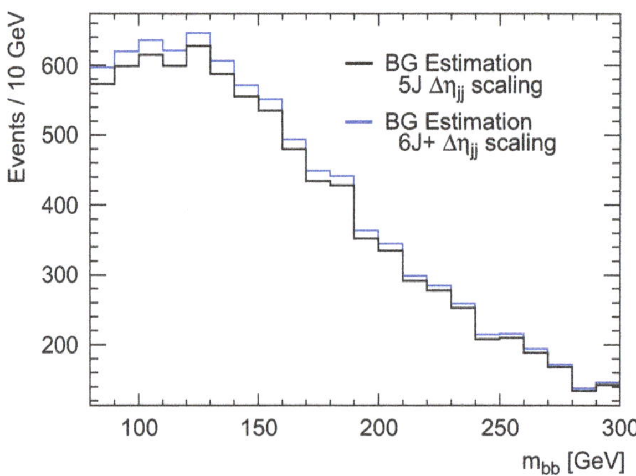

Fig. 7.2 Plots of m_{bb} using the background estimation with a tag region derived $\Delta\eta_{jj}$ linear correction (*black*) and a control region derived $\Delta\eta_{jj}$ correction (*blue*)

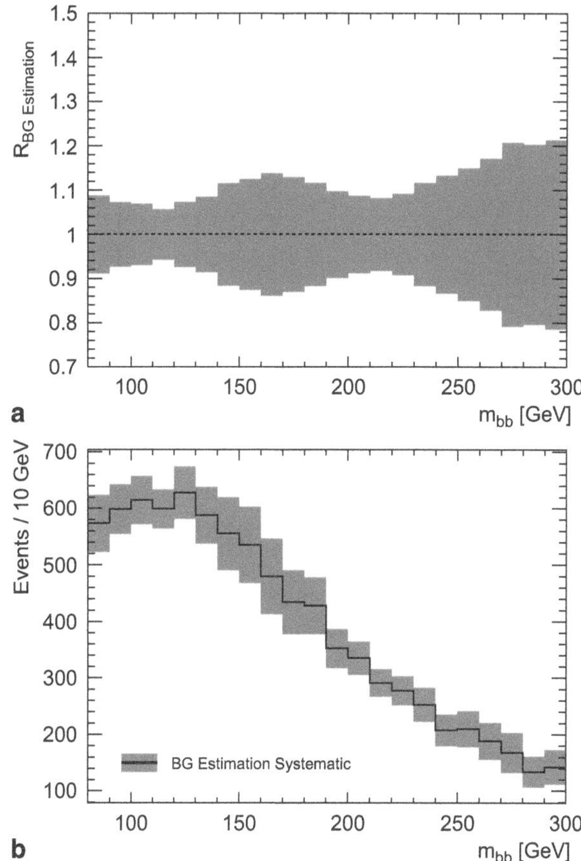

Fig. 7.3 MC derived uncertainty on the background estimation of m_{bb}: (**a**) relative uncertainty, R, derived from MC check ratio plot and (**b**) uncertainty applied to actual background estimated plot of m_{bb}

7.2 Monte Carlo Systematic Uncertainties

The MC samples used in this analysis all have fairly similar systematic uncertainties, with many mentioned briefly in previous chapters. They are described in more detail here, and the overall effect on acceptance and shape is summarized in Table 7.2.

7.2.1 b-Tagging Calibration

The major source of systematic uncertainty for MC samples is the b-tagging calibration systematic. As mentioned in Sect. 5.4, differences between data and MC samples

Table 7.2 Summary of the sources of uncertainty on the MC samples, averaged over all the signal and $t\bar{t}$ samples. The values correspond to the maximal effect on acceptance if each sources of uncertainty is varied by $\pm 1\sigma$. In the case of "Calibrations" and "Pile-up", the value are the result of adding several uncertainties in quadrature

Source of uncertainty	Effect on MC acceptance	
	Signal (averaged) (%)	$t\bar{t}$ (averaged) (%)
b-tagging calibration	21	20
JES		
Calibrations (various)	7	28
Pile-up (various)	2	6
Close-by jets	1	3
Flavour composition	10	15
Flavour response	5	9
Luminosity	1.8	1.8
Cross section	4.5	10

in b-tagging efficiencies are corrected using a scale factor, derived by per-jet weights that are functions of the jet p_T, η and flavour [1]. The scale factors have an uncertainty associated with them, that can range from 8 to 20 % per b-jet, and as high as 60 % for mistagged light-jets. To see the effect of the uncertainty on MC distributions, each scale factor is varied by its maximal associated uncertainty and propagated through the analysis. On average, this variation is about 23 % per bin in signal MC plots, as the example plot in Fig. 7.4 shows.

7.2.2 *Jet Energy Scale*

Another major source of uncertainty for MC samples is the Jet Energy Scale (JES) uncertainty. As mentioned in Sect. 4.4.1, the energy of jets in MC samples are corrected to match that of jets in data using η- and p_T-dependent scale factors, derived from various calibration algorithms. These corrections have intrinsic uncertainty as well, and ranges in value, as seen in Fig. 4.10. In general, on the signal MC samples used for this analysis, the JES uncertainty is on average 3 % for jets in the central region, and 5 % for forward jets.

The uncertainty is actually made up of 64 separate nuisance parameters, derived from uncertainties in the *in situ* calibrations, the η calibration, the flavour response and composition, as well as uncertainties from mis-measurements due to additional jets nearby, and pile-up. These uncertainties lead to both changes in the shape of kinematic distributions, as well as the overall acceptance. Therefore to derive their

[1] These scale factors (and their uncertainty) are calculated centrally by a flavour tagging working group at ATLAS.

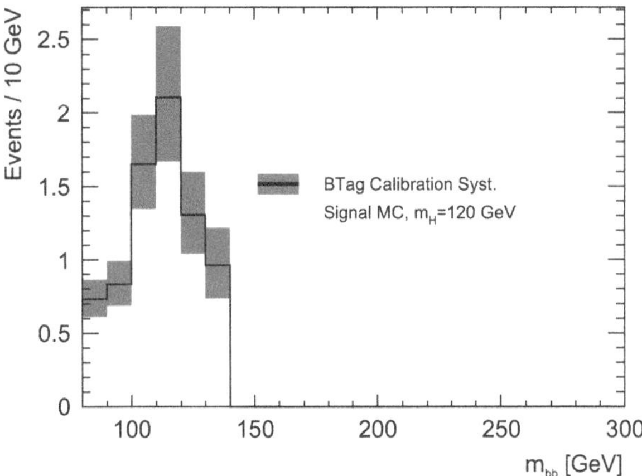

Fig. 7.4 Plot of m_{bb} using $m_H = 120$ GeV signal MC samples with b-tagging systematic uncertainty plotted

effect, for each nuisance parameter, the JES is varied by $\pm 1\sigma$ and propagated through the analysis. The result is upper and lower limits of the m_{bb} distribution for each of the parameters.

Of the 64 parameters, about half have no effect on the resulting m_{bb} distribution. Of the other half, the majority have a minimal effect, with maximum deviations per bin on the order of 4 to 10 %. There are some parameters, however, that lead to much greater deviations, the worst being the flavour composition uncertainty, which, has a maximum per bin deviation of 70 %. Example plots of m_{bb} with a JES nuisance parameter uncertainty that leads to maximal deviation, and a parameter uncertainty that leads to minimal deviation are shown in Fig. 7.5,

7.2.3 Luminosity

All MC samples are scaled to the accumulated luminosity of the data sample used (4.191 fb^{-1}). This luminosity has an intrinsic uncertainty of ± 1.8 % [67]. This is dominated by uncertainties in the calibration procedure to determine the luminosity.

7.2.4 Signal Specific Uncertainty

The cross sections and branching ratios mentioned in Table 4.1 have uncertainties associated to them. The uncertainty in the cross section is dominated by variations in the PDF used to derive their values ($\sim$2 %). Branching ratio uncertainties are dominated by variations in the QCD and electroweak corrections used to calculate

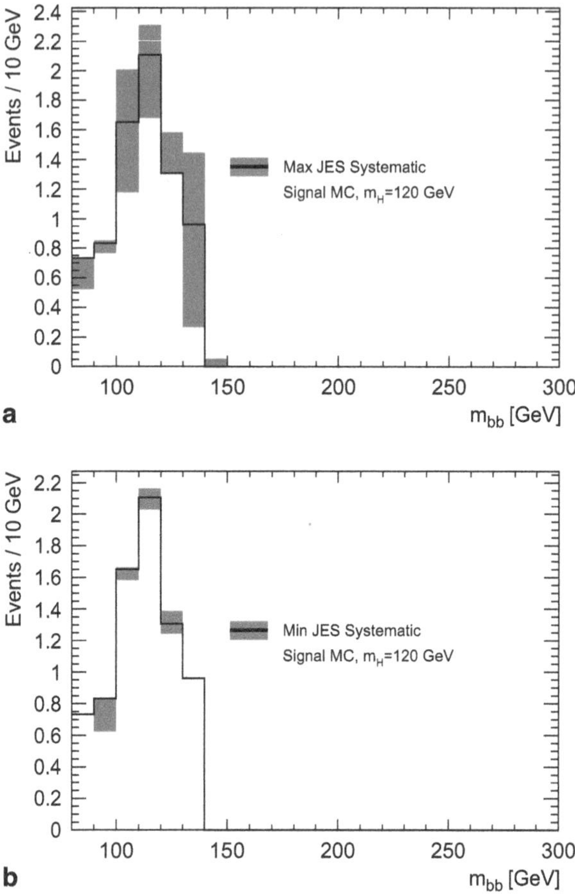

Fig. 7.5 Plots of m_{bb} distributions for $m_H = 120$ GeV signal MC samples with (**a**) the maximal JES uncertainty component applied (flavour composition) and (**b**) an example of a minimal JES uncertainty component applied (taken from Z+jets calibration)

them (∼2 %). These errors are added in quadrature, and an overall 4.5 % uncertainty is assigned to the signal cross section times branching ratio.

7.2.5 Background Specific Uncertainty

The other backgrounds considered in the limit calculation are the fully-hadronic and semi-leptonic $t\bar{t}$ samples, outlined in Table 4.5. The uncertainty in the cross section is taken to be a conservative 10 %, which is the upper limit of the uncertainty on these samples used in a similar analysis [68].

Chapter 8
Results and Discussion

The final m_{bb} distribution is plotted in Fig. 8.1. The plot shows the data points in the signal region (4J) and the background estimation discussed in Sect. 6.4. In addition, the plot shows the negligible contribution from the $t\bar{t}$ events (scaled to enhance its visibility) and the expected MC signal distribution (also scaled for enhanced visibility) for $m_H = 120$ GeV.

A 95 % CL exclusion limit calculation is performed, as per the method described in Appendix C, for Higgs masses of $m_H = 115, 120, 125, 130$ GeV using the invariant mass of the two b-tagged jets, m_{bb}, in the range of $80 < m_{bb} < 300$ GeV. The SM Higgs cross section can be excluded at the 95 % CL when the observed upper limit on μ (recall $\mu = \sigma/\sigma_{SM}$) dips below 1, as previously seen in Figs. 3.4 and 3.7.

The results are given in Table 8.1, and shown graphically in Fig. 8.2a. The observed 95 % CL upper limits on μ range from 14 to 19 over the mass range considered, whereas the expected upper limits range from 26 to 33 [1]. Although the observed values are systematically lower than the expected values, they are within the 2σ bands of the expected upper limits. This is likely due to the background estimation overshooting the data by a small amount, especially in the low m_{bb} region in Fig. 8.1.

All of the systematic uncertainties described in Chap. 7 were included in the CL limit calculation. To see the effect of these on the overall results, a similar limit calculation was conducted without the use of systematic uncertainties. The observed and expected values, plotted graphically in Fig. 8.2b, decrease only slightly, with 5 to 10 % improvement over the mass points considered. The observed limits are still systematically lower than the expected values, yet still within the $\pm 2\sigma$ bands of the expected 95 % CL upper limits.

[1] The observed upper limit is derived from the comparison of signal and background samples to the actual data samples, whereas the expected upper limit is the median expected upper limit for the background-only hypothesis derived from the signal and background samples.

E. Ouellette, *Search for the Higgs Boson in the Vector Boson Fusion Channel at the ATLAS Detector*, DOI 10.1007/978-3-319-13599-1_8

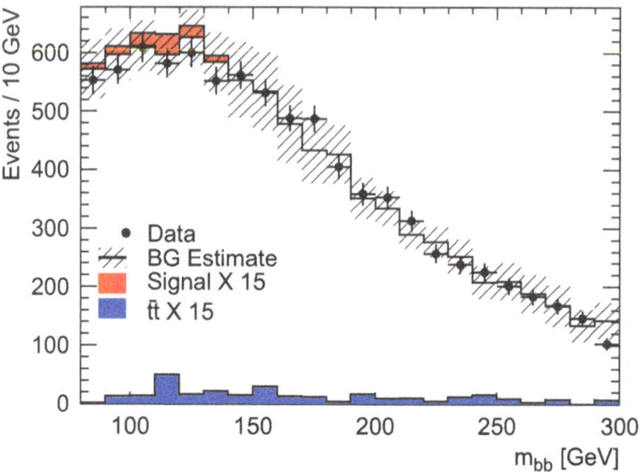

Fig. 8.1 Plot of m_{bb} with data points plotted, along with histogram stacks of $t\bar{t}$ MC samples (enhanced by factor of 15 for better visibility), background estimation (with shape uncertainty shaded in) and signal MC samples for $m_H = 120$ GeV (enhanced by factor of 15 for better visibility)

Table 8.1 Observed and expected 95 % CL upper limits on the SM Higgs cross section times branching ratio for the VBF $H^0 \to b\bar{b}$ channel for the tested Higgs masses. The $\pm 1\sigma$ and $\pm 2\sigma$ errors bands on the expected limits are also tabulated

Mass [GeV]	Observed	Expected	$+2\sigma$	$+1\sigma$	-1σ	-2σ
115	15.3	26.7	49.7	37.1	19.2	14.3
120	14.2	25.7	47.9	35.7	18.5	13.8
125	18.7	28.7	53.5	40.0	20.7	15.4
130	18.2	32.6	60.9	45.4	23.5	17.5

8.1 Comparison to Similar Analyses

The measurement finds no evidence for the Higgs boson. As a result, it cannot be used to further prove the discovery of the Higgs boson in the $H^0 \to b\bar{b}$ channel. As this is the first time this channel has been studied at ATLAS, it is useful to compare the results from this channel to those from other channels and other experiments.

ATLAS VH $H^0 \to b\bar{b}$

One obvious analysis with which to compare these results is the VH, $H^0 \to b\bar{b}$ analysis at ATLAS [69], which uses the VH production mechanism (see Fig. 3.1) instead of VBF. Using data collected in 2011, ATLAS was able to exclude from 2.5 to 5.5 times the SM Higgs cross section over the mass range of $110 < m_H < 130$ GeV, using a combination of the exclusion limits for three separate analyses: Z^0H^0 with $Z^0 \to l^+l^-$ ($l = e$ or μ), $W^{\pm}H^0$ with $W^{\pm} \to l^{\pm}\nu$, and Z^0H^0 with $Z \to \nu\nu$. If

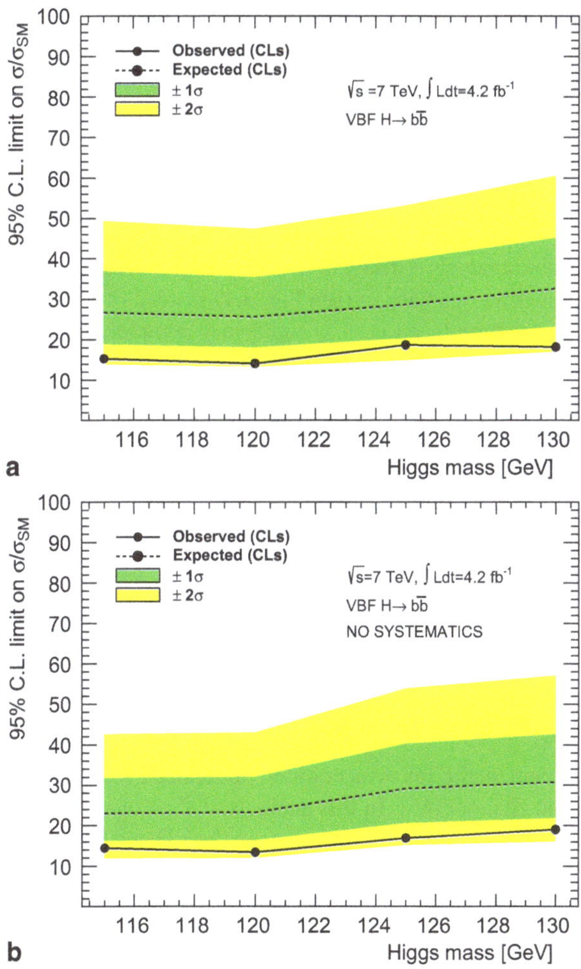

Fig. 8.2 Plots of the 95 % CL upper limit on the SM Higgs cross section times branching ratio for VBF $H^0 \rightarrow b\bar{b}$ including (**a**) all systematic uncertainties or (**b**) no systematic uncertainties, as a function of Higgs mass. The solid line shows the observed upper limits, whereas the dotted line shows the expected value with the $\pm 1\sigma$ and $\pm 2\sigma$ error bands on the expected value

one considers only one of these analyses, observed exclusion limits range from 7.7 to 14.4 times the SM cross section for the same mass range.

A direct comparison is difficult since the VH, $H^0 \rightarrow b\bar{b}$ channel had very different challenges to contend with. For one, it has the benefit of having high energy leptons in its final state or very large missing transverse energy. This simplifies the trigger used to collect data, as ATLAS had many dedicated low threshold single- and multi-lepton triggers in place, as well as missing transverse energy triggers with medium thresholds. The backgrounds were also reduced for these channels. For one of the analyses, total signal amounted to about 5 events using all of the 2011 data sample, with total background reaching 370 events in the invariant mass range of $80 < m_{bb} < 150$ GeV. That is nearly an order of magnitude smaller background compared to this channel, which has 7 signal events, for about 4200 background events in the same invariant mass range.

ATLAS ttH $\mathbf{H^0 \rightarrow b\bar{b}}$

Another comparable analysis is the ATLAS ttH, $H^0 \rightarrow b\bar{b}$ search channel [70], which uses the ttH production mechansim (see Fig. 3.1) instead of VBF. This channel has a far smaller cross section (see Fig. 3.2) and a very complex final state, with four or more jets, up to four b-tagged jets, missing transverse energy, and a reconstructed muon or electron. Similar to the VH analysis above, the leptons in the final state help in triggering events during data taking. The 95 % CL exclusion limits on the SM Higgs cross section is calculated using a combination of several jet multiplicity channels using 2011 data, and amounts to observed limits ranging from 9 to 16 over $115 < m_H < 130$ GeV, and expected limits ranging from 7 to 13. Although the upper limits are better than the limits obtained for VBF $H^0 \rightarrow b\bar{b}$, they are still complimentary to each other, as they both face similar challenges in their analyses, mainly in estimating background.

CDF VBF $\mathbf{H^0 \rightarrow b\bar{b}}$

The CDF experiment, at the Tevatron collider, published a search for the Higgs boson in the all hadronic channel [66]. In this analysis, 4 fb^{-1} of $\sqrt{s} = 1.96$ TeV proton-antiproton data is used to search for the Higgs in the VH and VBF production modes, concentrating on the $H^0 \rightarrow b\bar{b}$ decay channel, and the vector boson hadronic decay modes (for VH).

In contrast to this work, CDF used two different b-tagging algorithms, which lead to two different exclusion limits. The better of the two results for the VBF channel were observed (expected) 95 % CL upper limits ranging from 45 (47) to 70 (80), whereas the worst of the two were observed (expected) upper limits ranging from 80 (115) to 130 (200) over the Higgs mass range of $115 < m_H < 130$ GeV. Comparing exclusion limits, the ATLAS analysis had expected limits that were 40 to 60 % smaller and observed limits 70 % smaller over all mass points, compared to the better of the two CDF exclusion results. This is expected, since the centre-of-mass, $\sqrt{s}$, at the Tevatron was lower than at the LHC.

CDF used a similar background estimation method; however, a Neural Network technique was used to better separate between signal and background, using correlations between the variables m_{bb}, m_{jj}, and the η- and ϕ-moments (a measure of width) of the forward jets. Multivariate techniques tend to improve results; however, they must be very well understood. Besides, the goal of this thesis was to develop a first cut-based analysis for the VBF $H^0 \rightarrow b\bar{b}$ channel at ATLAS.

In general though, the results of the ATLAS analysis are consistent with those from the CDF analysis, as the exclusion limits are improved upon, as expected given the higher centre-of-mass of the LHC.

CMS VBF $H^0 \rightarrow b\bar{b}$

The CMS experiment (the other multi-purpose detector of the LHC) conducted a search using the VBF $H^0 \rightarrow b\bar{b}$ channel, using the 2012 $\sqrt{s} = 8$ TeV LHC data [71]. The analysis was different from this thesis, using fully simulated background MC samples instead of data driven background estimations, and using an artificial neural network to help discriminate signal and background. The observed and expected 95 % CL exclusion limits, using 19 fb^{-1} of 2012 data, range from 2 to 4 over $115 < m_H < 135$ GeV.

These improved results are due to several factors. Multivariate techniques typically discriminate between signal and background more effectively. The use of additional variables in the neural network utilizes correlations between variables, which is more difficult using a cut-based approach. The increased integrated luminosity and centre-of-mass also contributes to this improvement. The results from the CMS analysis suggest further progress is possible on the ATLAS analysis.

8.2 Discussion and Improvements

The results obtained in this analysis are first Higgs cross section exclusion limits using the VBF $H^0 \rightarrow b\bar{b}$ channel at ATLAS. The results are consistent with other analyses at ATLAS, and previous searches using this decay channel.

Additional work could be done to improve upon these results. The addition of the 2012 $\sqrt{s} = 8$ TeV dataset would increase the amount of data collected by a factor of four, as 22 fb^{-1} of data was collected during that time. The added data and the increased centre-of-mass would improve the statistics of the channel, thus pushing the exclusion limits down.

Fully simulated and well understood QCD background would also contribute to improving the results. This would prevent the need of a fully data driven method to estimate the background. Not only would this likely decrease the systematic uncertainties on the background, it would also help increase the signal efficiency. As mentioned in Sect. 6.4.5, 1b contributions of signal had to be subtracted from the background in order to avoid biases of an inflated signal contribution in the background estimation. This contribution is roughly 30 % of the signal in the 2b signal region, as shown in Fig. 6.9. This subtraction effectively weakens the signal by 30 %, and in turn, the exclusion limits are increased by the same amount. There was little to no signal in the tag region (5J) when developing the TRF; however, if MC samples were used, a loosening of selection criteria could provide an additional channel to use for discrimination between signal and background, which could be combined with the primary (4J) signal region.

When comparing this analysis to the CDF and CMS analyses, the use of multivariate techniques may improve the results of this analysis. Early in its development, there had been some preliminary work done to investigate the use of a multivariate approach for this channel. However, it was decided to concentrate on the cut-based approach due to two reasons. For one, there had not been an analysis developed for this channel, thus, it was deemed wise to develop a cut-based analysis first, in order to better understand the channel as a whole, and to set a baseline result for this channel. The second reason was due to the fact that early on, expertise lay in cut-based analyses for most other Higgs search channels at ATLAS, thus more support was available in developing an analysis strategy using cuts, rather than multivariate methods. This has evolved over time, and now many more analyses are being done at ATLAS using multivariate techniques. This analysis could be a starting point for future analyses using additional data and using multivariate techniques.

Appendix A
The Higgs Mechanism

The origin of the Higgs mechanism stems from the need to explain why the weak bosons and the fundamental fermions have mass. Without it, the weak theory is not locally gauge invariant, which causes it to be unrenormalizable. The Higgs field fixes this by having a non-zero vacuum expectation value, v.

This can be illustrated by looking at the spontaneous symmetry breaking of a $U(1)$ gauge field [7]. Consider the following Lagrangian of a complex scalar field, ϕ (which can be written in terms of scalar fields ϕ_1 and ϕ_2, $\phi = (\phi_1 + i\phi_2)/\sqrt{2}$):

$$\mathcal{L} = (\partial_\mu \phi^*)(\partial^\mu \phi) + \mu^2 \phi^* \phi - \lambda(\phi^* \phi)^2, \tag{A.1}$$

where ϕ has kinetic (first) and potential (last two) terms. If the constants μ and λ satisfy $\mu^2 > 0$ and $\lambda > 0$, then the ground state does not occur at $|\phi| = 0$, but instead is degenerate and occurs at any point where:

$$\phi^* \phi = \frac{1}{2}\left(\phi_1^2 + \phi_2^2\right) = \frac{\mu^2}{\lambda} \tag{A.2}$$

is satisfied. μ^2/λ is actually, by definition, the square of v. Since the Feynman calculus is perturbative, one must choose about which point to reformulate the Lagrangian, such that only fluctuations about the ground state are considered. To do this, the field ϕ can be rewritten in terms of the fields η_1 and η_2 from *any* minimum energy state (say $\phi_1 = v$ and $\phi_2 = 0$):

$$\phi = \sqrt{\frac{1}{2}}(v + \eta_1 + i\eta_2), \tag{A.3}$$

which, when put back into A.1, becomes:

$$\mathcal{L}' = \frac{1}{2}(\partial_\mu \eta_1)^2 + \frac{1}{2}(\partial_\mu \eta_2)^2 + \mu^2 \eta_1^2 + (\ldots), \tag{A.4}$$

where $(\ldots)$ represents constants and cubic and quartic interaction terms between η_1 and η_2. A.4 thus represents two scalar fields: one that is massive, η_1, the other massless, η_2. The appearance of this massless scalar field is due to the Goldstone

E. Ouellette, *Search for the Higgs Boson in the Vector Boson Fusion Channel at the ATLAS Detector*, DOI 10.1007/978-3-319-13599-1

theorem, which states that the spontaneous breaking of a continuous global symmetry will lead to massless scalar particles [1], called Goldstone bosons.

What is more interesting is when one considers the spontaneous breaking of a local gauge symmetry. For a local gauge transformation:

$$\phi \rightarrow e^{i\alpha(x)}\phi, \tag{A.5}$$

the partial derivatives in A.1 must be replaced with the covariant derivative;

$$D_\mu = \partial_\mu - ieA_\mu \tag{A.6}$$

where the gauge field A_μ transforms as:

$$A_\mu \rightarrow A_\mu + \frac{1}{e}\partial_\mu \alpha. \tag{A.7}$$

The locally gauge invariant Lagrangian then becomes:

$$\mathcal{L} = (\partial^\mu + ieA^\mu)\phi^*(\partial_\mu - ieA_\mu)\phi + \mu^2\phi^*\phi - \lambda(\phi^*\phi)^2 - \frac{1}{4}F_{\mu\nu}F^{\mu\nu}, \tag{A.8}$$

where $F_{\mu\nu} = \partial_\mu A_\nu - \partial_\nu A_\mu$. Now A.3 can still be used to rewrite this in terms of fluctuations about the ground state; however, a "miracle" occurs when local gauge invariance is used to choose an appropriate gauge. A.3 can be rewritten to lowest order as:

$$\phi \simeq \sqrt{\frac{1}{2}}(v + \eta_1)e^{i\eta_2/v} \tag{A.9}$$

and since A.8 is locally gauge invariant, A.5 can be used, with $\alpha = -\eta_2/v$, resulting in:

$$\phi' = \sqrt{\frac{1}{2}}(v + \eta_1) = \sqrt{\frac{1}{2}}(v + h) \tag{A.10}$$

where the field h is used instead of η_1. When put into A.8, this leads to the Lagrangian:

$$\mathcal{L} = \frac{1}{2}(\partial_\mu h)^2 - \lambda v^2 h^2 + evA_\mu^2 - \frac{1}{4}F_{\mu\nu}F^{\mu\nu} + (\ldots) \tag{A.11}$$

where $(\ldots)$ is the cubic and quartic interaction terms. This Lagrangian describes two massive particles: a vector gauge boson, A_μ and a massive scalar h. This massive scalar is the Higgs particle, and is responsible for the mass of the gauge boson. This is what is known as the Higgs mechanism.

Although this method only leads to one massive gauge boson, the same steps can be taken for an $SU(2)$ gauge symmetry, which leads to three massive vector

[1] The spontaneous breaking of a symmetry occurred here when the particular minimum energy state, $\phi_1 = v$, $\phi_2 = 0$, was chosen.

gauge bosons ($W^{\pm}$ and Z^0). Finally, the $SU(2)$ and $U(1)$ gauge symmetries can be combined together (see [8]), which leads to three massive vector gauge bosons (the weak bosons), one massless vector gauge bosons (the photon) and one massive scalar boson (the Higgs), which is consistent with the electroweak theory.

The Higgs mechanism can also be extended to fermions, generating fermionic masses in a similar fashion as the bosons. This is accomplished by adding gauge invariant Yukawa interactions between the Higgs field and the fermion field, for example:

$$\mathcal{L} = -G_f[\psi_L^\dagger \phi \psi_R + \psi_R^\dagger \phi_c \psi_L] \tag{A.12}$$

where G_f would be some constant (chosen such that $m_f = G_f v/\sqrt{2}$), and ψ_L and ψ_R would represent the left- and right-handed fermion. By choosing an appropriate gauge (just like the spontaneous symmetry breaking above), this Lagrangian becomes a sum of the fermion mass term and the fermion-fermion-Higgs coupling term:

$$\mathcal{L} = -m_f \bar{f} f - \frac{m_f}{v} \bar{f} f h. \tag{A.13}$$

This shows that the Higgs coupling to the fermions (the second term) is directly proportional to the mass of the fermion, as is evident in Fig. 3.3.

Appendix B
Triggers in 2012 and Beyond

Though data samples collected in 2012 were not used in this analysis, a trigger study was conducted to determine possible trigger signatures to be in place for this channel during the 2012 data taking period. The first issue to tackle was the L1 trigger rate. L1_4J10 could no longer be supported at the higher luminosities in 2012, since the rate increases exponentially, and can reach 50 kHz at a luminosity of 10^{34} cm^{-2} s^{-1}, nearly the limit of the L1 trigger. L1_4J15 was approved for 2012 data taking, and was thus used as the seed for 2012 triggers at HLT.

A study was conducted to see the effect of adding forward jets at L1. A similar study was done for 2011 data samples, though it didn't improve efficiencies since the L1 forward jets started at $|\eta| > 3.2$, and the forward jets of VBF events peak at $|\eta| \sim 2.5$. In 2012, there was a possibility of decreasing the L1 η threshold for forward jets to $|\eta| > 2.8$. Trigger signal efficiencies were computed for L1_4J15 and L1_3J15_FJ15, with the FJ15 having either a $|\eta| > 2.8$ or $|\eta| > 3.2$ threshold. The efficiency for the three jet trigger plus forward jet was comparable to four jet trigger if the looser η threshold was used. There was also little overlap between the two triggers, suggesting they would complement each other nicely during data taking.

Very little flexibility is available at L1, as trigger algorithms must be very fast in order to stay within the limits of data taking at that level. However, in the HLT, there is more information and time to make a more refined decision. A significant fraction of the signal is lost when cutting on the p_T of each of the four leading jets. Cutting on quantities such as m_{jj} or $\Delta\eta_{jj}$ would be much more beneficial, as less signal would be lost at the trigger level.

Further rank optimized offline selection criteria of $\Delta\eta_{jj} > 3.0$ and $m_{jj} >$ 662 GeV where determined using data and MC samples. A study was conducted to determine the effect on signal of applying these selection criteria online. Two L1 triggers were considered as baseline, seed triggers: L1_4J15 and L1_3J15_FJ15 (with the lower $|\eta|$ threshold mentioned above). After applying loose preselection criteria (four jets with $p_T > 25$ GeV, two of which are b-tagged), L1_4J15 has a higher overall signal efficiency (2.8 %) compared to L1_3J15_FJ15 (2.2 %). However, when comparing plots of m_{jj} vs. $\Delta\eta_{jj}$, shown in Fig. B.1, L1_3J15_FJ15 clearly has a higher concentration of events at greater values of m_{jj} and $\Delta\eta_{jj}$ (as expected, these variables are correlated). This is also shown clearly in Fig. B.2.

E. Ouellette, *Search for the Higgs Boson in the Vector Boson Fusion Channel at the ATLAS Detector*, DOI 10.1007/978-3-319-13599-1

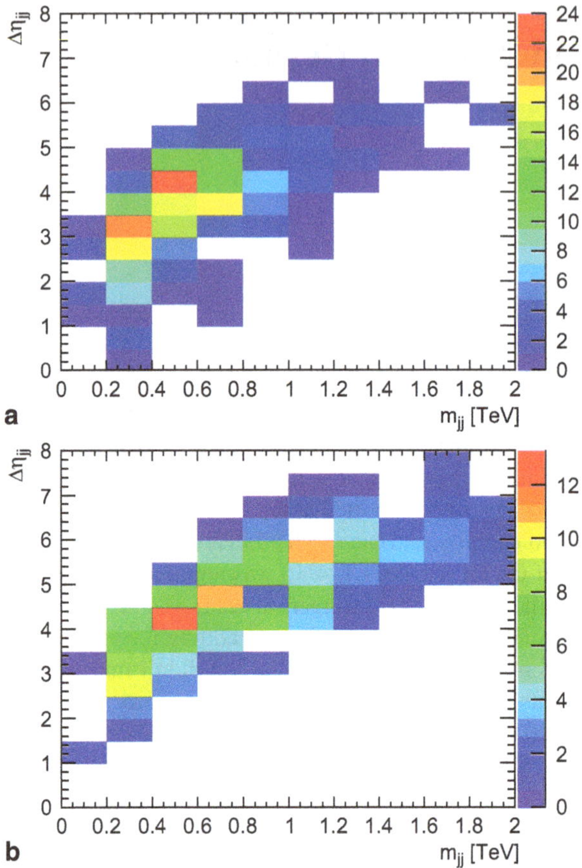

Fig. B.1 Plots of m_{jj} vs. $\Delta\eta_{jj}$ for $m_H = 120$ GeV signal MC samples after requiring (**a**) L1_4j15 or (**b**) L1_3J15_FJ15, four jets with $p_T > 25$ GeV and two jets that satisfy offline b-tagging. The z axis is number of MC events

To stay within the limits of the offline selection criteria, online trigger criteria of $m_{jj} > 400$ GeV and $\Delta\eta_{jj} > 3.0$ were proposed. As expected, L1_3J15_FJ15 was less affected, with an 89 % relative selection efficiency, compared to L1_4J15, with 76 %. Though even after the cut, L1_4J15 had a higher overall signal efficiency, 2.1 %, slightly greater than L1_3J15_FJ15, with a 1.9 % overall efficiency.

Approximate background rates were calculated using MC samples, with 10^{34} cm^{-2} s^{-1} as a benchmark luminosity. Similar to the signal MC samples, L1_4J15 had a slightly greater rate than L1_3J15_FJ15 (630 Hz compared to 540 Hz). After requiring two offline b-tagged jets, the rate for L1_4J15 was about 4 Hz, which would be too high and would not be accepted at HLT unprescaled, whereas L1_3J15_FJ15 had a more acceptable rate of 1.5 Hz. Finally after applying the m_{jj} and $\Delta\eta$ criteria "online", both trigger seeds had rates on the order of 1 Hz,

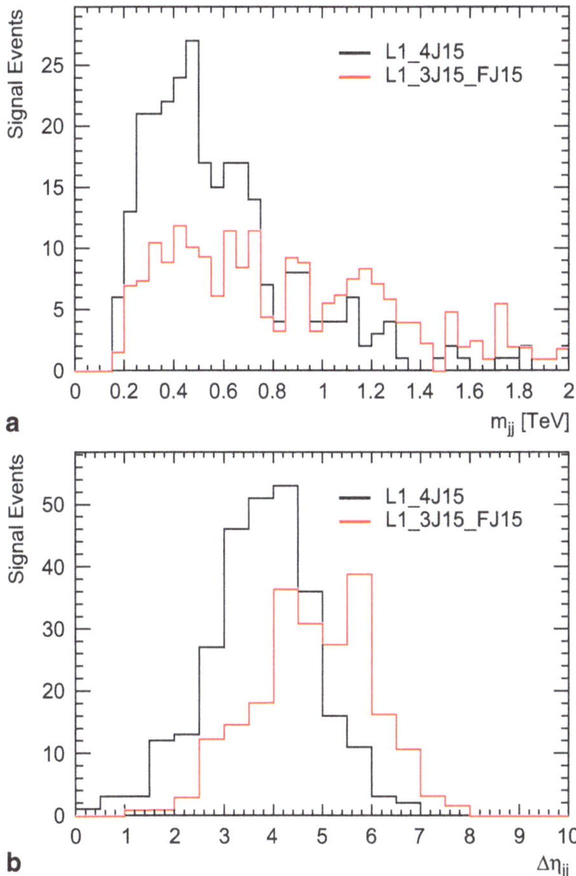

Fig. B.2 Plots of (**a**) m_{jj} and (**b**) $\Delta\eta_{jj}$ for $m_H = 120$ GeV signal MC samples comparing L1_4J15 and L1_3J15_FJ15 triggers after applying cuts on number of jets (≥ 4 with $p_T > 25$ GeV) and number of b-jets (≥ 2 offline b-tagged jets). The y axis is number of MC events

which is in the acceptable range of rate per trigger signature. As a result, a so called VBF trigger item was implemented in late 2012, which cuts loosely on m_{jj} and $\Delta\eta_{jj}$ for certain L1 multijet trigger seeds.

Appendix C
Statistical Tests for Particle Physics

The discovery of the Higgs boson, as outlined in Chap. 3, required an extremely large dataset, since the search for the Higgs has been akin to "finding a needle in a haystack". It is therefore essential that very thorough statistical tests be used to quantify the search, and subsequent discovery of the Higgs. ATLAS uses a frequentist method to set limits on observation, discovery and exclusion [72].

C.1 Test Statistic for Particle Searches

When searching for new processes in particle physics experiments, it is necessary to quantify the significance of an observed signal. This is often done by means of a p-value: the probability, under some hypothesis, of finding data of equal or greater compatibility with the prediction of the hypothesis. The hypothesis can be regarded as excluded if the observed p-value is below a specified threshold. A p-value is often converted to an equivalent significance, Z, defined such that a Gaussian distributed variable found Z standard deviations above its mean has upper-tail probability equal to p, illustrated in Fig. C.1a. These are related by:

$$Z = \Phi^{-1}(1 - p) \tag{C.1}$$

where Φ^{-1} is the inverse of the cumulative distribution of the standard Gaussian, plotted in Fig. C.1b. The standard for discovery (or the rejection of the background-only hypothesis) within the particle physics community is a significance of $Z = 5$, which corresponds to $p = 2.87 \times 10^{-7}$, whereas the standard for excluding a signal hypothesis is a threshold of $p = 0.05$ (i.e. a 95 % Confidence Level (CL)), which corresponds to $Z = 1.64$.

The most common procedure to establish discovery or exclusion is based on a frequentist significance test using a likelihood ratio as a test statistic. Consider an experiment where for each selected event, one measures a kinematic variable, x, to search for a signal (the invariant mass of two b-jets for example). A histogram, $\mathbf{n} = (n_1, \ldots, n_N)$, is then constructed from selected events, where the expectation

E. Ouellette, *Search for the Higgs Boson in the Vector Boson Fusion Channel at the ATLAS Detector*, DOI 10.1007/978-3-319-13599-1

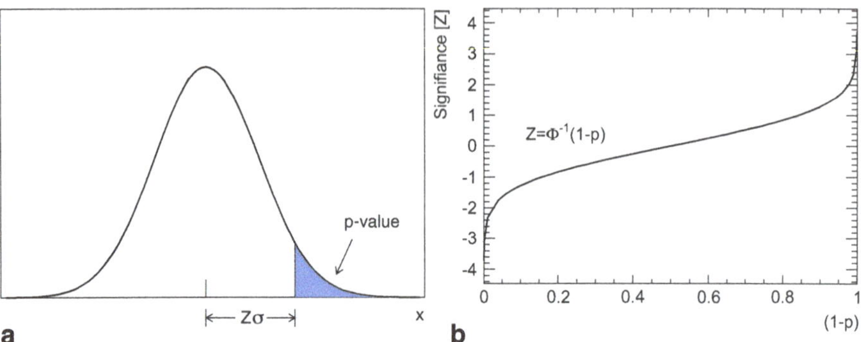

Fig. C.1 Plots of (**a**) the correspondence between Z and p-value for a gaussian distribution and (**b**) the inverse cumulative distribution of the standard Gaussian, Φ^{-1}, as a function of $(1-p)$. Taken from [72]

value of n_i can be written:

$$E[n_i] = \mu s_i + b_i \tag{C.2}$$

1 where the mean number of entries in the ith bin from signal and background are:

$$s_i = s_{tot} \int_{\text{bin } i} f_s(x; \boldsymbol{\theta}_s)\, dx, \tag{C.3}$$

$$b_i = b_{tot} \int_{\text{bin } i} f_b(x; \boldsymbol{\theta}_b)\, dx. \tag{C.4}$$

The parameter μ corresponds to the strength of the signal process, with $\mu = 0$ corresponding to a background-only hypothesis, and $\mu = 1$ being the nominal signal hypothesis. The functions $f_s(x; \boldsymbol{\theta}_s)$ and $f_b(x; \boldsymbol{\theta}_b)$ are the probability densities of the variable x for both signal and background events, where $\boldsymbol{\theta}_s$ and $\boldsymbol{\theta}_b$ represent the parameters that characterize the shape of the functions. These functions are integrated over the bin i, and multiplied by s_{tot} or b_{tot}, the total mean number of signal and background events respectively. Both the signal and background models contain nuisance parameters whose values are not taken as known *a priori*, but rather must be fitted from the data. These are denoted as $\boldsymbol{\theta} = (\boldsymbol{\theta}_s, \boldsymbol{\theta}_b, b_{tot})$. It is assumed that the signal normalization, s_{tot}, is fixed to the value predicted by the nominal signal model, whereas μ acts as an adjustable strength parameter.

Each bin of the variable is assumed to follow Poisson statistics, therefore the likelihood function is the product of Poisson probabilities over all bins:

$$L(\mu, \boldsymbol{\theta}) = \prod_{j=1}^{N} \frac{(\mu s_j + b_j)^{n_j}}{n_j!} e^{-(\mu s_j + b_j)}, \tag{C.5}$$

then to test a hypothesized value of μ, a profile likelihood ratio is calculated:

$$\lambda(\mu) = \frac{L(\mu, \hat{\hat{\boldsymbol{\theta}}})}{L(\hat{\mu}, \hat{\boldsymbol{\theta}})}, \tag{C.6}$$

where $\hat{\hat{\boldsymbol{\theta}}}$ denotes the value of $\boldsymbol{\theta}$ that maximizes L for the specified value of μ (conditional maximum-likelihood estimator of $\boldsymbol{\theta}$). The quantities $\hat{\mu}$ and $\hat{\boldsymbol{\theta}}$ are the values of μ and $\boldsymbol{\theta}$ that maximizes L (the unconditional maximum-likelihood). The presence of the nuisance parameters broadens the profile likelihood as a function of μ, compared to what one would expect if their values were fixed. The profile likelihood, λ, satisfies $0 \leq \lambda \leq 1$, with values near 1 implying good agreement between data and the expected background and signal models for the hypothesized value of μ.

In most analyses, the signal process contribution is assumed to be non-negative, which implies $\mu > 0$. However, as will be seen below, $\hat{\mu}$ will be modelled as a Gaussian distributed variable, and therefore must be allowed to take on negative values.

C.1.1 Test Statistic for Upper Limits

For establishing upper limits on μ, ATLAS uses the following test statistic:

$$q_\mu = \begin{cases} -2\ln\lambda(\mu) & \hat{\mu} \leq \mu, \\ 0 & \hat{\mu} > \mu, \end{cases} \tag{C.7}$$

where $\lambda(\mu)$ is defined in Eq. C.6. Setting $q_\mu = 0$ for $\hat{\mu} > \mu$ is due to the fact that data satisfying this condition would not represent less compatibility with μ than the data obtained, thus one is not trying to reject this region. From the definition, higher values of q_μ represent greater incompatibility between data and the hypothesized value of μ.

The level of agreement between data and a hypothesized value of μ is quantified using a p-value, defined as:

$$p_\mu = \int_{q_{\mu,\text{obs}}}^{\infty} f(q_\mu|\mu)dq_\mu, \tag{C.8}$$

where $f(q_\mu|\mu)$ is the probability density function of q_μ assuming a value of μ.

To set a 95 % CL upper limit, the CL_S method [73] is used, defined as:

$$CL_S \equiv \frac{p_{s+b}}{1-p_b} < \alpha \tag{C.9}$$

where one finds the value of μ for which its p-value, p_{s+b}, divided by $1 - p_b$ (p_b is the p-value for the $\mu = 0$ case), is less than some threshold value, often $\alpha = 0.05$ (i.e. 95 % CL). This method is preferred over simply using p_{s+b}, since it is much more conservative in cases where there is little sensitivity to the signal model, though it results in higher upper limits. In the case of large sensitivity to the signal model, there is little difference between CL_S and p_{s+b}.

C.1.2 Approximating Sampling Distributions

In order to determine the p-value of a hypothesis in Eqs. C.8 and C.9, one needs to determine the distribution $f(q_\mu|\mu)$. It is also important to determine the distribution of $f(q_\mu|\mu')$, where $\mu \neq \mu'$, in order to determine the expected significance if the data corresponds to a strength parameter, μ', different from the one being tested, μ. This distribution has been approximated to be:

$$-2\ln\lambda(\mu) = \frac{(\mu-\hat{\mu})^2}{\sigma^2} + \mathcal{O}(1/\sqrt{N}), \tag{C.10}$$

where $\hat{\mu}$ is a Gaussian distribution with mean μ' and standard deviation σ, and N is the data sample size. The standard deviation σ can be obtained from a covariance matrix of the estimators for all parameters (i.e. the nuisance parameters $\boldsymbol{\theta}$). Assuming a large enough data sample to ignore the second term above, the probability density function for Eq. C.10 follows a non-central χ^2 distribution. This simplifies even further to a normal χ^2 distribution in the special case of $\mu' = \mu$.

This can then be used to approximate the distribution for $f(q_\mu|\mu)$ in Eq. C.8:

$$f(q_\mu|\mu') = \Phi\left(\frac{\mu'-\mu}{\sigma}\right)\delta(q_\mu) + \frac{1}{2\sqrt{2\pi q_\mu}}\exp\left[-\frac{1}{2}\left(\sqrt{q_\mu} - \frac{\mu-\mu'}{\sigma}\right)^2\right], \tag{C.11}$$

which simplifies to a half-χ^2 distribution in the special case $\mu = \mu'$. To find the upper limit on μ, one can find the largest value of μ that satisfies $p_\mu \leq \alpha$, for some value of α, i.e. $\alpha = 0.05$ for the 95 % CL.

C.1.3 Experimental Sensitivity

The sensitivity of the experiment is characterized not by the significance obtained from a single data set, but rather the expected, or median significance with which one can reject different values of μ. This is illustrated in Fig. C.2, which shows example distributions of $f(q_\mu|\mu)$ and $f(q_\mu|\mu')$. The median value of $f(q_\mu|\mu')$ can be used to calculate the median value of the p-value, assuming μ'. For setting exclusion limits, one uses $\mu' = 0$ as the assumed hypothesis, rejecting nonzero values of μ from the calculated p-value.

For the purpose of determining the median expected significance, one replaces the ensemble of simulated background and signal data by a single representative data set, referred to as the "Asimov" data set. This is derived by estimating each parameter of $\boldsymbol{\theta}$ to be the value that maximizes the likelihood function with respect to that parameter. Theses values are those implied by the "assumed" distribution of the data, and are estimated from the MC model using a large data sample.

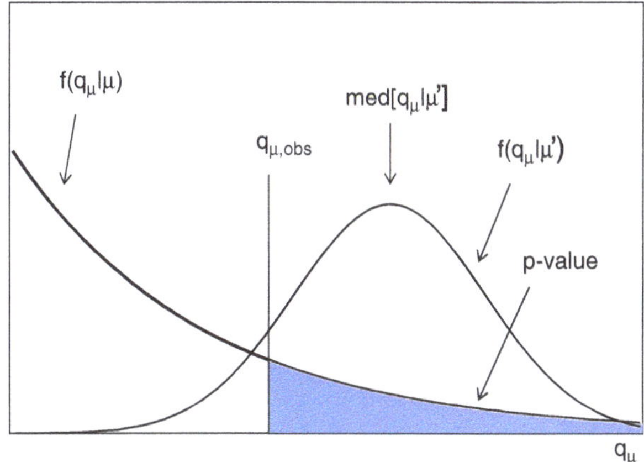

Fig. C.2 Illustration of the probability distribution function of q_μ for both a strength parameter μ and a different value μ'. The plotted p-value demonstrates the experiments sensitivity as it gives the median p-value assuming μ'. Taken from [72]

This Asimov data can in turn be used to evaluate an "Asimov likelihood", L_A, which is then used in the profile likelihood ratio, $-2\ln\lambda_A(\mu)$. This leads to a $q_{\mu,A}$, which can be replaced in each instance of q_μ in Eq. C.11, to be used to find the median upper limit on μ assuming $\mu' = 0$.

The actual data contains statistical fluctuations, and thus the observed significance is not in general equal to the median. It is thus useful to determine by how much the significance may vary, given the expected fluctuations in the data. Fortunately, the approximate sampling distributions can predict how the significance is expected to vary under the assumption of a given signal strength. Error bands for the median significance are then often quoted, which corresponds to $\pm N\sigma$ variations about $\hat{\mu}$ (which is of course a Gaussian distributed variable, about μ').

Fortunately all these methods have been implemented into one central C++ class library called RooStats [74], based on the ROOT and RootFit [75] packages. The framework allows for minimizations (for example to obtain the estimators in the profile likelihood ratio), which are conducted using the Minuit [76] package.

Bibliography

1. Glashow, S. Partial symmetries of weak interactions. Nucl Phys. 1961;22:579–88.
2. Weinberg, S. A model of leptons. Phys Rev Lett. 1967;19:1264–66.
3. Salam A. Weak and electromagnetic interactions. Conf Proc. 1968;C680519:367–377.
4. MissMJ. Standard model of elementary particles. Accessed 2013-10-23 via Wikipedia. Creative Commons License.
5. Particle Data Group Collaboration, Beringer J, et al. Review of particle physics (RPP). Phys Rev. 2012;D86:010001.
6. Griffiths DJ. Introduction to elementary particles. WILEY-VCH: Weinheim; 2004.
7. Halzen F, Martin AD. Quarks and leptons: an introductory course in modern particle physics. Wiley: New York; 1984.
8. Aitchison IJR, Hey AJG. Gauge theories in particle physics: a practical introduction. 3rd ed. Graduate student series in physics. IOP: Bristol; 2004.
9. Englert F,Brout R. Broken symmetry and the mass of gauge vector mesons. Phys Rev Lett. 1964;13:321–323.
10. Higgs PW. Broken symmetries, massless particles and gauge fields. Phys Lett. 1964;12: 132–133.
11. Higgs PW. Broken symmetries and the masses of gauge bosons. Phys Rev Lett. 1964;13: 508–509.
12. Guralnik G, Hagen C, Kibble T. Global conservation laws and massless particles. Phys Rev Lett. 1964;13:585–587.
13. Higgs PW. Spontaneous symmetry breakdown without massless bosons. Phys Rev. 1966;145:1156–1163.
14. Kibble T. Symmetry breaking in non-abelian gauge theories. Phys Rev. 1967;155:1554–1561.
15. Das AK, Ferbel T. Introduction to nuclear and particle physics. New York; 1995.
16. Martin A, Stirling W, Thorne R, et al. Parton distributions for the LHC. Eur Phys J. 2009;C63:189–285. arXiv:0901.0002 [hep-ph].
17. Campbell JM, Huston J, Stirling W. Hard interactions of quarks and gluons: a primer for LHC physics. Rept Prog Phys. 2007;70:89. arXiv:0611148 [hep-ph].
18. Ellis J, Gaillard MK, Nanopoulos DV. A historical profile of the higgs boson. arXiv:1201.6045 [hep-ph].
19. Andari N, Assamagan K, Bourgaux A, et al. Higgs production cross sections and decay branching ratios, Tech. Rep. ATL-COM-PHYS-2010-046., CERN, Geneva, 2010.
20. Dittmaier S, Mariotti C, Passarino G, et al. Handbook of LHC higgs cross sections: 1. inclusive observables. Geneva: CERN; 2011.
21. Barbieri R, Ericson TEO. Evidence against the existence of a low mass scalar boson from neutron–nucleus scattering, Phys Lett. 1975;B57:270.

E. Ouellette, *Search for the Higgs Boson in the Vector Boson Fusion Channel at the ATLAS Detector*, DOI 10.1007/978-3-319-13599-1

22. LEP Working Group for Higgs boson searches, ALEPH Collaboration, DELPHI Collaboration, L3 Collaboration, OPAL Collaboration, Barate R, et al. Search for the standard model Higgs boson at LEP. Phys Lett. 2003;B565:61–75. arXiv:0306033 [hep-ex].
23. CDF Collaboration, D0 Collaboration, Aaltonen T, et al. Higgs Boson studies at the Tevatron. Phys Rev. 2013;D88:052014. arXiv:1303.6346 [hep-ex].
24. ATLAS Collaboration, Aad G, et al., Search for the Standard Model Higgs boson in the diphoton decay channel with 4.9 fb^{-1} of pp collisions at $\sqrt{s} = 7$ TeV with ATLAS, Phys Rev Lett. 2012:108:111803. arXiv:1202.1414 [hep-ex].
25. ATLAS Collaboration, Aad G, et al. Measurements of the properties of the Higgs-like boson in the two photon decay channel with the ATLAS detector using 25 fb^{-1} of proton-proton collision data Tech Rep. ATLAS-CONF-2013-012, CERN, Geneva, Mar, 2013.
26. ATLAS Collaboration, Aad G, et al. Search for the standard model Higgs boson in the decay channel $H \rightarrow$ ZZ(*) $\rightarrow 4\ell$ with 4.8 fb^{-1} of pp collision data at $\sqrt{s} = 7$ TeV with ATLAS. Phys Lett. 2012;B710:383–402. arXiv:1202.1415 [hep-ex].
27. ATLAS Collaboration, G. Aad et al. Measurements of the properties of the Higgs-like boson in the four lepton decay channel with the ATLAS detector using 25 fb^{-1} of proton-proton collision data, Tech Rep. ATLAS-CONF-2013-013, CERN, Geneva, Mar, 2013.
28. ATLAS Collaboration, G. Aad et al. Update of the $H \rightarrow WW^{(*)} \rightarrow e\nu\mu\nu$ Analysis with 13 fb^{-1} of $\sqrt{s} = 8$ TeV data collected with the ATLAS detector Tech Rep. ATLAS-CONF-2012-158, CERN, Geneva, Nov, 2012.
29. ATLAS Collaboration, Aad G, et al. Search for the Standard Model Higgs boson in $H \rightarrow \tau\tau$ decays in proton-proton collisions with the ATLAS detector. Tech Rep. ATLAS-CONF-2012-160, CERN, Geneva, Nov, 2012.
30. ATLAS Collaboration, Aad G, et al. Search for the standard model Higgs boson produced in association with a vector boson and decaying to bottom quarks with the ATLAS detector. Tech Rep. ATLAS-CONF-2012-161, CERN, Geneva, Nov, 2012.
31. ATLAS Collaboration Aad G et al. An update of combined measurements of the new Higgs-like boson with high mass resolution channels Tech Rep. ATLAS-CONF-2012-170, CERN, Geneva, Dec, 2012.
32. CMS Collaboration, Chatrchyan S, et al. Observation of a new boson at a mass of 125 GeV with the CMS experiment at the LHC. Phys Lett. 2012;B716: 30–61. arXiv:1207.7235 [hep-ex].
33. CMS Collaboration. Properties of the Higgs-like boson in the decay H to ZZ to 4l in pp collisions at $\sqrt{s}$ =7 and 8 TeV. Tech Rep. CMS-PAS-HIG-13-002, CERN, Geneva, 2013.
34. CMS Collaboration. Updated measurements of the Higgs boson at 125 GeV in the two photon decay channel. Tech Rep. CMS-PAS-HIG-13-001, CERN, Geneva, 2013.
35. CMS Collaboration. Combination of standard model Higgs boson searches and measurements of the properties of the new boson with a mass near 125 GeV. Tech Rep. CMS-PAS-HIG-13-005, CERN, Geneva, 2013.
36. ATLAS Collaboration, Aad G, et al. Evidence for the spin-0 nature of the Higgs boson using ATLAS data. Phys Lett. 2013 ;B726: 120–144. arXiv:1307.1432 [hep-ex].
37. CMS Collaboration, Chatrchyan S, et al. Study of the mass and spin-parity of the Higgs Boson candidate via its decays to Z Boson Pairs. Phys Rev Lett. 2012;110:081803. 25 p. arXiv:1212.6639 [hep-ex].
38. Hankele V , Klamke G , Zeppenfeld D, et al. Anomalous Higgs boson couplings in vector boson fusion at the CERN LHC, Phys Rev. 2006;D74:095001. arXiv:0609075 [hep-ph].
39. Bruening OS , Collier P , Lebrun P, et al. LHC Design Report. CERN, Geneva, 2004.
40. Horvath A. "LHC." Accessed 2013-09-30 via Wikipedia. Creative Commons License.
41. ATLAS Collaboration, Aad A, et al. The ATLAS experiment at the CERN large hadron collider. JInstrum. 2008;3(08):S08003.
42. Unno Y. ATLAS silicon microstrip semiconductor tracker (SCT). Nucl Instrum Methods Phys Res, A 2009;453 (1–2):109–20.
43. ATLAS TRT Collaboration, Abat E, et al. The ATLAS TRT Barrel Detector. JInstrum. 2008;3 no.(02):P02014.

44. ATLAS TRT Collaboration, Abat E, et al. The ATLAS TRT end-cap detectors. J Instrum. 2008;3(10):P10003.
45. ATLAS Collaboration, Aad A, et al., Atlas computing: technical design report. Geneva: CERN; 2005.
46. ROOT: an object-oriented data analysis framework: users guide. CERN, Geneva, 2007.
47. Golling T, Hayward H , Onyisi P, et al. The ATLAS data quality defect database system. Eur Phys J. 1960;C72(2012). arXiv:1110.6119 [physics.ins-det].
48. ATLAS Collaboration, Aad G, et al. https://twiki.cern.ch/twiki/bin/view/AtlasPublic/LuminositvPublicResults. Accessed 2013 Oct 21.
49. ATLAS Collaboration, Aad G, et al. Expected performance of the ATLAS Experiment—detector, Trigger Phys. arXiv:0901.0512 [hep-ex].
50. Cacciari M, Salam GP, Soyez G. The anti-k(t) jet clustering algorithm. JHEP. 2008;0804:063, arXiv:0802.1189 [hep-ph].
51. ATLAS Collaboration, Aad G, et al. Jet energy scale and its systematic uncertainty in Proton-Proton Collisions at $\sqrt{s}$ = 7 TeV with ATLAS 2011 Data, Tech. Rep. ATLAS-CONF-2013-004. CERN, Geneva, Jan, 2013.
52. Bousson N, Vacavant L, Bai Y, et al. Commissioning of the ATLAS high-performance b-tagging algorithms in the 7 TeV collision data, Tech. Rep. ATLAS-COM-CONF-2011-110, CERN, Geneva, June, 2011.
53. ATLAS Collaboration, Aad G, et al. Measurement of the b-tag efficiency in a sample of jets containing muons with 5fb^{-1} of Data from the ATLAS Detector, Tech. Rep. ATLAS-CONF-2012-043, CERN, Geneva, March, 2012.
54. Sjostrand T, Mrenna S, Skands PZ. PYTHIA 6.4 physics and manual. JHEP. 2006;0605:026. arXiv:0603175 [hep-ph].
55. Mangano ML, Moretti M, Piccinini F, et al. ALPGEN, a generator for hard multiparton processes in hadronic collisions. JHEP. 2003;07:001. arXiv:0206293 [hep-ph].
56. Corcella G, Knowles I, Marchesini G, et al. HERWIG 6 : an event generator for hadron emission reactions with interfering gluons (including supersymmetric processes). JHEP. 2001;1 :010.
57. Butterworth J, Forshaw JR, Seymour M. Multiparton interactions in photoproduction at HERA. Z Phys. 1996;C72:637–46. arXiv:9601371 [hep-ph].
58. Frixione S, Stoeckli F, Torrielli P, et al. The MC@NLO 4.0 Event Generator. arXiv:1010.0819 [hep-ph].
59. GEANT4 Collaboration, Agostinelli S, et al. GEANT4: a simulation toolkit. Nucl Instrum Meth. 2003;A506 (2003):250–303.
60. Shibata A, Bosman M, Hawkings R, et al. Understanding monte carlo generators for top physics, Tech. Rep. ATL-COM-PHYS-2009-334, CERN, Geneva, Jan, 2009.
61. Mangano M, Moretti M, Piccinini F et al. b anti-b final states in Higgs production via weak boson fusion at the LHC. Phys Lett. 2003;B556:50–60. arXiv:0210261 [hep-ph].
62. Hansson Adrian P. The ATLAS b-Jet trigger, Tech. Rep. ATL-COM-DAQ-2011-114, CERN, Geneva, Oct, 2011.
63. Coccaro A, Hansson P. Weighting method for b-Jet trigger heavy flavor tagging calibration, Tech. Rep. ATL-COM-DAQ-2011-144, CERN, Geneva, Nov, 2011.
64. ATLAS Collaboration, Aad G, et al. Selection of jets produced in proton-proton collisions with the ATLAS detector using 2011 data, Tech. Rep. ATLAS-CONF-2012-020, CERN, Geneva, March, 2012.
65. ATLAS Collaboration, Aad G, et al. Performance of the ATLAS jet trigger in the early $\sqrt{s}$=7 TeV data, Tech. Rep. ATLAS-CONF-2010-094, CERN, Geneva, Oct, 2010.
66. CDF Collaboration, Aaltonen T, et al. Search for the Higgs boson in the all-hadronic final state using the CDF II detector. Phys Rev. 2011;D84:052010. arXiv:1102.0024 [hep-ex].
67. ATLAS Collaboration, Aad G, et al. Improved luminosity determination in pp collisions at $\sqrt{s}$ = 7 TeV using the ATLAS detector at the LHC. Eur Phys J. 2013;C73:2518. arXiv:1302.4393 [hep-ex].

68. Ahmadov F, Allbrooke BMM, Anjos N et al. Searches for a Higgs boson decaying to a b-quark pair with the ATLAS detector at the LHC, Tech. Rep. ATL-COM-PHYS-2010-929, CERN, Geneva, Nov, 2010.
69. ATLAS Collaboration, Aad G, et al. Search for the Standard Model Higgs Boson produced in association with a vector boson and decaying to a b-quark pair with the ATLAS detector. Phys Lett. 2012;B718:369–390. arXiv:1207.0210 [hep-ex].
70. ATLAS Collaboration, Aad G, et al. Search for the Standard Model Higgs Boson produced in association with top quarks in proton-proton collisions at $\sqrt{s} = 7$ TeV using the ATLAS detector, Tech. Rep. ATLAS-CONF-2012-135, CERN, Geneva, Sept, 2012.
71. CMS Collaboration. Higgs to bb in the VBF channel, Tech. Rep. CMS-PAS-HIG-13-011, CERN, Geneva, 2013.
72. Cowan G, Cranmer K, Gross E, et al. Asymptotic formulae for likelihood-based tests of new physics. Eur Phys J. 2011;C71:1554. arXiv:1007.1727, [physics.data-an].
73. Read AL. Presentation of search results: the CL s technique. J Phys G Nucl Part Phys. 2002; 28(10):2693.
74. Moneta L, Belasco K, Cranmer KS, et al. The roostats project. PoS ACAT2010. 2010;057. arXiv:1009.1003 [physics.data-an].
75. Verkerke W, Kirkby DP. The roofit toolkit for data modeling. eConf. 2003;C0303241: MOLT007. arXiv:0306116 [physics].
76. James F, Roos M. Minuit: a system for function minimization and analysis of the parameter errors and correlations. Comput Phys Commun. 1975;10:343–367.

Zeitfracht Medien GmbH
Ferdinand-Jühlke-Straße 7
99095 Erfurt, Deutschland
produktsicherheit@kolibri360.de